PRAISE FOR

FINAL JEOPARDY

"Now that the *Jeopardy!* game has been played and won, this background will help readers understand some of the rather dismissive remarks already appearing in the press from leading AI researchers." —*New York Journal of Books*

"The book's hook is the *Jeopardy!* contest, but its potential readership is considerably more various than the show's fans." —*Booklist*

"Baker skillfully weaves the two threads of the story together, and the book contains many passages that make the reader not only assess what they think but how they think, and how they have absorbed and stored the knowledge they possess. It's books like this that remind us there is still so much we don't understand about our own brains, and that the journey of discovery has only just begun." —*CultureMob*

"Plenty of observers have weighed in on Watson . . . few have done it better than Stephen Baker." —*Bloomberg*

"[The] primary focus here is on the compelling story of Watson's creation and education. . . . This is a thought-provoking view of one of IBM's major contributions to the computing field." —*Library Journal*

FINAL JEOPARDY

BOOKS BY STEPHEN BAKER

The Numerati

Final Jeopardy

STEPHEN BAKER

FINAL JEOPARDY

THE STORY OF WATSON,

THE COMPUTER THAT WILL

TRANSFORM OUR WORLD

MARINER BOOKS

HOUGHTON MIFFLIN HARCOURT

BOSTON • NEW YORK

www.hmhbooks.com

Library of Congress Cataloging-in-Publication Data
Baker, Stephen, date.
Final Jeopardy: the story of Watson, the computer that will transform our world /
Stephen Baker.
p. cm.
ISBN 978-0-547-48316-0
ISBN 978-0-547-74719-4 (pbk.)
1. Natural language processing (Computer science) 2. Semantic computing.
3. Artificial intelligence. 4. Database management. 5. Watson (Computer)
6. *Jeopardy* (Television program)
I. Title.
QA76.9.N38B35 2011
006.3—dc22
2010051653

Book design by Melissa Lotfy

Printed in the United States of America
DOC 10 9 8 7 6 5 4 3 2 1

To Sally and Jack, my fact-hungry sister and son, who each introduced me to Jeopardy! *at a different stage of my life*

Contents

FINAL JEOPARDY

Introduction

WATSON PAUSED. The closest thing it had to a face, a glowing orb on a flat-panel screen, turned from forest green to a dark shade of blue. Filaments of yellow and red streamed steadily across it, like the paths of jets circumnavigating the globe. This pattern represented a state of quiet anticipation as the supercomputer awaited the next clue. It was a September morning in 2010 at IBM Research, in the hills north of New York City, and the computer, known as Watson, was annihilating two humans, both champion players, in practice rounds of *Jeopardy!* Within months, it would be playing the game on national television in a million-dollar man vs. machine match against two of *Jeopardy*'s all-time greats.

As Todd Crain, an actor and the host of these test games, started to read the next clue, the filaments on Watson's display began to jag and tremble. Watson was thinking—or coming as close to it as a computer could. The $1,600 clue, in the category The Eyes Have It, read: "This facial ware made Israel's Moshe Dayan instantly recognizable worldwide."

The three players—two human and one electronic—could

read the words as soon as they appeared on the big *Jeopardy* board. But they had to wait for Crain to read the entire clue before buzzing. That was the rule. As the host pronounced the last word, a light would signal that contestants could buzz. The first to hit the button could win $1,600 with the right answer—or lose the same amount with a wrong one. (In these test matches, they played with funny money.)

This pause for reading gave Watson three or four seconds to hunt down the answer. The first step was to figure out what the clue meant. One of its programs promptly picked apart the grammar of the sentence, identifying the verbs, objects, and key words. In another section, research focused on Moshe Dayan. Was this a person? A place in Israel? Perhaps a holy site? Names like John and Maria would signal a person. But Moshe was more puzzling.

During these seconds, Watson's cognitive apparatus— 2,208 computer processors working in concert—mounted a massive research operation through thousands of documents around Moshe Dayan and his signature facial ware. After a second or so, different programs, or algorithms, began to suggest hundreds of possible answers. To us, many of them would look like wild guesses. Some were phrases that Dayan had uttered, others were references to his military campaigns and facts about Israel. Still others cited various articles of his clothing. At this point, the computer launched its second stage of analysis, figuring out which response, if any, merited its confidence. It proceeded to check and recheck facts, making sure that Moshe Dayan was indeed a person, an Israeli, and that the answer referred to something he wore on his face.

A person looking at Watson's frantic and repetitive labors might conclude that the player was unsure of itself, laughably short on common sense, and scandalously wasteful of computing resources. This was all true. Watson barked up every tree from every conceivable angle. The pattern on its screen during this process, circles exploding into little stars, provided only a hint of the industrial-scale computing at work. In a room behind the podium, visible through a horizontal window, Watson's computers churned, and the fans cooling them roared. This time, its three seconds of exertion paid off. Watson came up with a response, sending a signal to a mechanical device on the podium. It was the size of a large aspirin bottle with a clear plastic covering. Inside was a *Jeopardy* buzzer. About one one-hundredth of a second later, a metal finger inside this contraption shot downward, pressing the button.

Justin Bernbach, a thirty-eight-year-old airline lobbyist from Brooklyn, stood to Watson's left. He had pocketed $155,000 while winning seven straight *Jeopardy* matches in 2009. Unlike Watson, Bernbach understood the sentence. He knew precisely who Moshe Dayan was as soon as he saw the clue, and he carried an image of the Israeli leader in his mind. He gripped the buzzer in his fist and frantically pressed it four or five times as the light came on.

But Watson had arrived first.

"Watson?" said Crain.

The computer's amiable male voice arranged the answer, as *Jeopardy* demands, in the form of a question: "What is eye patch?"

"Very good," Crain said. "An eye patch on his left eye. Choose again, Watson."

Bernbach slumped at his podium. This match with the machine wasn't going well.

It was going magnificently for David Ferrucci. As the chief scientist of the team developing the *Jeopardy* computer, Ferrucci was feeling vindicated. Only three years earlier, the suggestion that a computer might match wits and word skills with human champions in *Jeopardy* sparked opposition bordering on ridicule in the halls of IBM Research. And the final goal of the venture, a nationally televised match against two *Jeopardy* legends, Ken Jennings and Brad Rutter, seemed risky to some, a bit déclassé to others. *Jeopardy*, a television show, appeared to lack the timeless cachet of chess, which IBM computers had mastered a decade earlier.

Nonetheless, Ferrucci and his team went ahead and built their machine. Months earlier, it had fared well in a set of test matches. But the games revealed flaws in the machine's logic and game strategy. It was a good player, but to beat Jennings and Rutter, who would be jousting for a million-dollar top prize, it would have to be great. So they had worked long hours over the summer to revamp Watson. This September event was the coming-out party for Watson 2.0. It was the first of fifty-six test matches against a higher level of competitor: people, like Justin Bernbach, who had won enough matches to compete in *Jeopardy*'s Tournament of Champions.

In these early matches, Watson was having its way with them. Ferrucci, monitoring the matches from a crowded observation booth, was all smiles. Keen to promote its *Jeopardy* phenom, IBM's advertising agency, Ogilvy & Mather, had hired a film crew to follow Ferrucci's team and capture the drama of this opening round of championship matches. The observation room was packed with cameras. Microphones on long booms recorded the back-and-forth of engineers as they discussed algorithms and Watson's response time, known as latency. Ferrucci, wearing a mike on his lapel, gave a blow-by-blow commentary as Watson, on the other side of the glass, strutted its new and smarter self.

It was almost as if Watson, like a person giddy with hubris, was primed for a fall. The computer certainly had its weaknesses. Even when functioning smoothly, it would make its share of wacky mistakes. Right before the lunch break, one clue asked about "the inspiration for this title object in a novel and a 1957 movie [which] actually spanned the Mae Khlung." Now, it would be reasonable for a computer to miss "The Bridge over the River Kwai," especially since the actual river has a different name. Perhaps Watson had trouble understanding the sentence, which was convoluted at best. But how did the computer land on its outlandish response, "What is Kafka?" Ferrucci didn't know. Those things happened, and Watson still won the two morning matches.

It was after lunch that things deteriorated. Bernbach, so frustrated in the morning, started to beat Watson to the buzzer. Meanwhile, the computer was making risky bets and flubbing entire categories of clues. Defeat, which had seemed so remote in the morning, was now just one lost bet away. It came in the fourth match. Watson was winning by $4,000 when it stumbled on this Final Jeopardy clue: "On Feb. 8, 2010, the headline in a major newspaper in this city read: 'Amen! After 43 years, our prayers are answered.'" Watson missed the reference to the previous day's Super Bowl, won by the New Orleans Saints. It bet $23,000 on Chicago. Bernbach also botched the clue, guessing New York. But he bet less than Watson, which made him the first person to defeat the revamped machine. He pumped his fist.

In the sixth and last match of the day, Watson trailed Bernbach, $16,200 to $21,000. The computer landed on a Daily Double in the category Colleges and Universities, which meant it could bet everything it had on nailing the clue. A $5,000 bet would have brought it into a tie with Bernbach. A larger bet, while risky, could have catapulted the computer toward victory. "I'll take five," Watson said.

Five. Not $5,000, not $500. Five measly dollars of funny money. The engineers in the observation booth were stunned. But they kept quieter than usual; the cameras were rolling.

Then Watson crashed. It occurred at some point between placing that lowly bet and attempting to answer a clue about the first Catholic college in Washington, D.C. Watson's "front end," its voice and avatar, was waiting for its thousands of processors, or "back end," to deliver an answer. It received nothing. Anticipating such a situation, the engineers had prepared set phrases. "Sorry," Watson said, reciting one of them, "I'm stumped." Its avatar displayed a dark blue circle with a single filament orbiting mournfully in the Antarctic latitudes.

What to do? Everyone had ideas. Maybe they should finish the game with an older version of Watson. Or perhaps they could hook it up to another up-to-date version of the program at the company's Hawthorne labs, six miles down the road. But some worried that a remote connection would slow Watson's response time, causing it to lose more often on the buzz. In the end, as often happens with computers, a reboot brought the hulking *Jeopardy* machine back to life. But Ferrucci and his team got an all-too-vivid reminder that their *Jeopardy* player, even as it prepared for a debut on national television, could go haywire or shut down at any moment. When Watson was lifted to the podium, facing banks of cameras and lights, it was anybody's guess how it would perform. Watson, it was clear, had a frighteningly broad repertoire.

Only four years earlier, in 2006, Watson was a prohibitive long shot, not just to win at *Jeopardy* but even to be built. For more than a year, the head of IBM Research, a physicist named Paul Horn, had been pressing a number of teams at the company to pursue a *Jeopardy* machine. The way he saw it, IBM had triumphed in 1997 with its chess challenge. The company's machine, Deep Blue, had defeated the reigning world champion, Garry Kasparov. This burnished IBM's reputation among the global computing elite while demonstrating to the world that computers could rival human beings in certain domains associated with intelligence.

That triumph left IBM's top executives hungry for an encore. Horn felt the pressure. But what could the researchers get a computer to do? Deep Blue had rifled through millions of scenarios per second, calculated probabilities, and made winning moves. It had aced applied math. But it had skipped the far more complex domain of words. This, Horn thought, was where the next challenge would be. Far beyond the sixty-four squares on a chess board, the next computer should charge into the vast expanse of human language and knowledge. For the test, Horn settled on *Jeopardy*, which debuted in 1964 and now attracted some nine million viewers every weeknight. It was the closest thing in the United States to a knowledge franchise. "People associated it with intelligence," Horn later said.

There was one small problem. For months, he couldn't get any takers. *Jeopardy*, with its puns and strangely phrased clues, seemed too hard for a computer. IBM was already building machines to answer questions, and their performance, in speed and precision, came nowhere close to that of even a moderately informed person. How could the next machine grow so much smarter?

And while researchers regarded the challenge as daunting, many people, Horn knew, saw it precisely the other way. Answering questions? Didn't Google already do that?

Horn eventually enticed David Ferrucci and his team to pursue his vision. Ferrucci, then in his midforties, wore a dark brown beard wrapped around his mouth and wire-rimmed glasses. An expert in Artificial Intelligence (AI), he had a native New Yorker's gift of the gab and an openness, even about his own life, that was at times jolting. ("I have a growing list of potentially mortal diseases," he said years later. "People order an MRI a week for me.") But he also had a wide and ranging intellect. Early in his tenure at IBM he and a friend tried, in their spare time, to teach a machine to write fiction by itself. They trained it for various literary themes, from love to betrayal, and they named it Brutus, for Julius Caesar's traitorous comrade. Ferrucci was comfortable talking about everything from the details of computational linguistics to the evolution of life on earth and the nature of human thought. This made him an ideal ambassador for a *Jeopardy* machine. After all, his project would raise a broad range of issues, and fears, about the role of brainy machines in society. Would they compete for jobs? Could they establish their own agendas, like the infamous computer HAL, in *2001: A Space Odyssey,* and take control? What was the future of knowledge and intelligence, and how would brains and machines divvy up the cognitive work? Ferrucci was always ready with an opinion. At the same time, he could address the strategic questions—how these machines would fit into hundreds of businesses, and why the project he was working on, as he saw it, went far beyond Google.

The Google question was his starting point; until people understood that his machine was not just a souped-up search engine, the project made little sense. For certain types of questions, Ferrucci said, a search engine could come up with answers. These were simple sentences with concrete results, what he and his team called factoids. For example: "What is the tallest mountain in Africa?" A search engine would pick out the three key words from that sentence and in a fraction of a second suggest Kenya's 19,340-foot-high Kilimanjaro. This worked, Ferrucci said, for about 30 percent of *Jeopardy* questions. But performance at that low level would condemn Watson to defeat at the hands of human amateurs.

A *Jeopardy* machine would have to master far thornier questions. Just as important, it would have to judge its level of confidence in an answer. Google's algorithms delivered users to the statistically most likely outposts of the Web and left it to the readers to find the answers. "A search engine doesn't know that it understood the question and that the content is right," Ferrucci said. But a *Jeopardy* machine would have to find answers and then decide for itself if they were worth betting on. Without this judgment, the machine would never know when to buzz. It would require complex analysis to develop this "confidence."

Was it worth it? Didn't it make sense for machines to hunt through mountains of data and for people to rely on their exquisitely engineered brains to handle the final judgments? This seemed like a reasonable division of labor. After all, processing language and spotting answers come easily to humans and are so hard for machines.

But what if machines could take the next step? What if they could go beyond locating bits and pieces of information and help us to understand it? "I think there are 1.4 million books on sale online," Ferrucci said one afternoon. He was sharing a bottle of his own wine, a Shiraz blend that he'd cooked up in the winery attached to his kitchen in the northern suburbs of New York. He was in an expansive mood, which led him to carry out energetic dialogues with himself, asking questions and answering them emphatically. "You can

probably fit all the books that are on sale on about two tera-
bytes that you can buy at OfficeMax for a couple hundred
dollars. You get every book. Every. Single. Book. Now what
do you do? You can't read them all! What I want the computer
to do," he went on, "is to read them for me and tell me what
they're about, and answer my questions about them. I want
this for *all* information. I want machines to read, understand,
summarize, describe the themes, and do the analysis so that I
can take advantage of all the knowledge that's out there. We
humans need help. I know I do!"

Before building a *Jeopardy* machine, Ferrucci and his team
had to carry this vision one step further: They had to make a
case that a market existed outside the rarefied world of *Jeopardy* for advanced question-answering technology. IBM's biggest division, after all, was Global Services, which included
one of the world's largest consultancies. It sold technical and
strategic advice to corporations all over the world. Could
the consultants bundle this technology into their offerings?
Would this type of machine soon be popping up in offices
and answering customers' questions on the phone?

Ferrucci envisioned a *Jeopardy* machine spawning a host
of specialized know-it-alls. With the right training, a technology that could understand everyday language and retrieve answers in a matter of seconds could fit just about anywhere.
Its first job would likely be in call centers. It could answer
tax questions, provide details about bus schedules, ask about
the symptoms of a laptop on the fritz and walk a customer
through a software update. That stuff was obvious. But there
were plenty of other jobs. Consider publicly traded companies, Ferrucci said. They had to comply with a dizzying assortment of rules and regulations, everything from leaks of
inside information in e-mails to the timely disclosure of earn-

ings surprises or product failures to regulators and investors. A machine with Watson's skills could stay on top of these compliance matters, pointing to possible infractions and answering questions posed in ordinary English. A law firm could call on such a machine to track down the legal precedent for every imaginable crime, complaint, or trademark.

Perhaps the most intriguing opportunity was in medicine. While IBM was creating the *Jeopardy* machine, one of the top medical shows on television featured a nasty genius named Gregory House. In the beginning of most episodes a character would collapse, tumbling to the ground during a dance performance, a lovers' spat, or a kindergarten class. Each one suffered from a different set of symptoms, many of them gruesome. In the course of the following hour, amid the medical team's social and sexual dramas, House and his colleagues would review the patient's worsening condition. There had to be a pattern. Who could find it and match it to a disease, ideally before the patient died? Drawing from their own experience, the doctors each mastered a diverse set of data. The challenge was to correlate that information to the ever-changing list of symptoms on the white board in House's office. Toward the end of the show, House would often notice some detail—perhaps a lyric in a song or an unlikely bruise. And that would lead his magnificent mind straight to a case he remembered or a research paper he'd read about bee stings or tribal rites in New Guinea. By the end of the show, the patient was headed toward recovery.

An advanced question-answering machine could serve as a bionic Dr. House. Unlike humans, it could stay on top of the tens of thousands of medical research papers published every year. And, just as in *Jeopardy*, it could come up with lists of potential answers, or diagnoses, for each patient's ills. It could also direct doctors toward the evidence it had considered and provide its reasoning. The machine, lacking common sense, would be far from perfect. Just as the *Jeopardy* computer was certain to botch a fair number of clues, the diagnoses coming from a digital Dr. House would sometimes be silly. So people would still run the show, but they'd be assisted by a powerful analytical tool.

In those early days, only a handful of researchers took part in the *Jeopardy* project at IBM. They could fit easily into Ferrucci's office at the research center in Hawthorne, New York, about thirty-five miles north of New York City (and a fifteen-minute drive from corporate headquarters, in Armonk). But to build a knowledge machine, Ferrucci knew, would require extensive research and development. In a sense, a *Jeopardy* machine would represent an entire section of the human brain. To build it, he would need specialists in many aspects of cognition. Some would be experts in language, others in the retrieval of information. Some would attempt to program the machine with judgment, writing algorithms to steer it toward answers. Others would guide it in so-called machine learning, so that it could train itself to pick the most statistically promising combinations of words and pay more attention to trustworthy sources. Experts in hardware, meanwhile, would have to build a massive computer, or a network of them, to process all of this work. Assembling these efforts on a three-year timetable amounted to a daunting management challenge. The cost of failure would be humiliation, for both the researchers and their company.

Other complications came from the West Coast, specifically the Robert Young building on the Sony lot in Culver City, a neighborhood just south of Hollywood. Unlike chess, a treasure we all share, the *Jeopardy* franchise belonged to Sony

Pictures Entertainment, an arm of the Japanese consumer electronics giant. The *Jeopardy* executives, led by a canny negotiator named Harry Friedman, weren't about to let IBM use their golden franchise and their millions of viewers on its own terms. Over the years, the two companies jousted over the terms of the game, the placement of logos, access to stars such as Ken Jennings and Brad Rutter, and the writing of *Jeopardy* clues. They even haggled over the computer's speed on the buzzer and whether IBM should take measures to slow it to a human level. These disagreements echoed until the eve of the match. At one point, only months before the showdown, *Jeopardy*'s executives appeared to be on the verge of pulling the plug on the entire venture. That would have left IBM's answering computer, the product of three intense years of research, scrounging for another game to play. This particular disagreement was resolved. But the often conflicting dictates of promotion, branding, science, and entertainment forged a fragile and uneasy alliance.

The *Jeopardy* project also faced harsh critics within IBM's own scientific community. This was to be expected in a field—Artificial Intelligence—where the different beliefs about knowledge, intelligence, and the primacy of the human brain bordered on the theological. How could there be any consensus in a discipline so vast? While researchers in one lab laboriously taught machines the various meanings of the verb "to do," futurists just down the hall insisted that computers would outrace human intelligence in a couple of decades, controlling the species. Beyond its myriad approaches and outlooks, the field could be divided into two camps, idealists and pragmatists. The idealists debated the nature of intelligence and aspired to build computers that could think conceptually, like human beings, perhaps surpassing us. The pragmatists created machines to carry out tasks. Ferrucci, who had promised to have a television-ready computer by early 2011, fell firmly into the second camp—and his team attracted plenty of barbs for it. The *Jeopardy* machine would sidestep the complex architecture of the brain and contrive to answer questions without truly understanding them. "It's just another gimmick," said Sajit Rao, a professor in computer science at MIT who's attempting to teach computers to conceptualize forty-eight different verbs. "It's not addressing any fundamental problems." But as Ferrucci countered, teaching a machine to answer complex questions on a broad range of subjects would represent a notable advance, whatever the method.

IBM's computer would indeed come to answer a dizzying variety of questions—and would raise one of its own. With machines like this in our future, what do we need to store in our own heads? This question, of course, has been recurring since the dawn of the Internet, the arrival of the calculator, and even earlier. With each advance, people have made internal adjustments and assigned ever larger quantities of memory, math, geography, and more to manmade tools. It makes sense. Why not use the resources at hand? In the coming age, it seems, forgoing an effective answering tool will be like volunteering for a lobotomy.

In a sense, many of us living through this information revolution share something with the medieval monks who were ambushed by the last one. They spent years of their lives memorizing sacred texts that would soon be spilling off newfangled printing presses. They could have saved lots of time, and presumably freed up loads of capacity, by archiving those texts on shelves. (No need to discuss here whether the monks were eager for "free time," a concept dangerously close to

Sloth, the fourth of the Seven Deadly Sins.) In the same way, much of the knowledge we have stuffed into our heads over the years has been rendered superfluous by new machinery.

So what does this say about Ken Jennings and Brad Rutter, the humans preparing to wage cognitive war with Watson? Are they relics? Sure, they might win this round. But the long-term prognosis is grim. Garry Kasparov, the chess master who fell to IBM's Deep Blue, recently wrote that the golden age of man-machine battles in chess lasted from 1994 to 2004. Before that decade, machines were too dumb; after it, the roles were reversed. While knowledge tools, including Watson, relentlessly advance, our flesh-and-blood brains, some argue, have stayed more or less the same for forty thousand years, treading evolutionary water from the Cro-Magnon cave painters to Quentin Tarantino.

A few decades ago, know-it-alls like Ken Jennings seemed to be the model of human intelligence. They aced exams. They had dozens of facts at their fingertips. In one quiz show that predated *Jeopardy*, *College Bowl*, teams of the brainiest students would battle one another for the honor of their universities. Later in life, people turned to them in boardrooms, university halls, and cocktail parties for answers. Public education has been designed, in large part, to equip millions with a ready supply of factual answers. But if Watson can top them, what is this kind of intelligence worth?

Physical strength has suffered a similar downgrade. Not so long ago, a man with superhuman strength played a valuable role in society. He was a formidable soldier. When villagers needed boulders moved or metal bent, he got the call. After the invention of steam engines and hydraulic pumps, however, archetypal strongmen were shunted to jobs outside the productive economy. They turned to bending metal in circuses or playing noseguard on Sunday afternoons. For many of us, physical strength, once so vital, has become little more than a fashion statement. Modern males now display muscles as mating attire, much the way peacocks fan their otherwise useless feathers.

It would be all too easy to dismiss human foes of the IBM machine as cognitive versions of circus strongmen: trivia wunderkinds. But from the very beginning, Ferrucci saw that the game required far more than the simple regurgitation of facts. It involved strategy, decision making, pattern recognition, and a knack for nuance in the language of the clues. As the computer grew from a whimsical idea into a *Jeopardy* behemoth, it underwent an entire education, triumphing in some areas, floundering in others. Its struggles, whether in untangling language or grappling with abstract ideas, highlighted the areas in which humans maintain an edge. It is in the story of Watson's development that we catch a glimpse of the future of human as well as machine intelligence.

The secret is wrapped up in the nature of knowledge itself. What is it? For humans, knowledge is an entire universe, a welter of sensations and memories, desires, facts, skills, songs and images, words, hopes, fears, and regrets, not to mention love. But for those hoping to build intelligent machines, it has to be simpler. Broadly speaking, it falls into three categories: sensory input, ideas, and symbols. Consider the color blue. Sensory perception is the raw material of knowledge. It's something that computers and people alike can perceive, each in their own fashion. Now think of the word "sky." Those three letters are a symbol for the biggest piece of blue in our world. Computers can handle such symbols. They can find references to "sky" in documents and, when programmed, correlate it with others, such as "blue," "clouds," and "heaven." A computer can master both sensory data and symbols. It can count, categorize, search, and store them. But how about this snippet from Lord Byron: "Friendship is love without his wings." That sentence represents the third realm of knowledge: ideas. How can a machine make sense of them? In these early years of the twenty-first century, ideas remain the dominion of people—and the frontier for thinking machines.

David Ferrucci's mission was to explore that frontier. Like many in his profession, Ferrucci grew up watching *Star Trek* on television. The characters on the show, humans and pointy-eared Vulcans alike, spoke to their computer as if it were one of them. No formatting was necessary, no key words, no programming language. They spoke English. The computer understood the meaning and context of the questions. It consulted vast databases and came back with an immediate answer. True, it might not produce original ideas. But it was an extravagantly well-informed shipmate. That was the computer Ferrucci wanted to build.

As he served the last drops of his wine, Ferrucci was talking about the world he was busy creating, one in which people and their machines often appeared to switch roles. He didn't know, he said, whether engineers would ever be able to "create a sentient being." But when he looked at his fellow humans through the eyes of a computer scientist, he saw patterns of behaviors that often appeared to be programmed. He mentioned the zombielike commutes, the retreat to the same chair, the hand reaching for the TV remote, and the near-identical routines, from toothbrushing to feeding the animals. "It's more interesting," he said, "when humans delve inside themselves and say, 'Why am I doing this? And why is it relevant and important to be human?'" His machine would nudge people toward that line of inquiry. Even with an avatar for a face and a robotic voice, the *Jeopardy* machine would invite comparisons to the other two contestants on the stage. This was inevitable. And whether it won or lost on a winter evening in 2011, the computer might lead millions of spectators to reflect on the nature, and probe the potential, of their own humanity.

1

The Germ of the *Jeopardy* Machine

THE *JEOPARDY* MACHINE's birthplace—if a computer can stake such a claim—was the sprawling headquarters of the global research division named after its flesh-and-blood ancestor, IBM's founder, Thomas J. Watson. In 1957, when IBM presided over the rest of the infant computer industry, the company cleared woods on a hill in Yorktown Heights, New York, about forty miles north of midtown Manhattan, and hired the Finnish-American architect Eero Saarinen to design a lab. If computing was the future, as seemed inevitable, it was on this hill that a good part of it would be dreamed up, modeled mathematically, and prototyped. Saarinen was a natural choice to express this sparkling future in glass and rock. A year earlier, he had designed the winged TWA Terminal for the new Idlewild Airport (later called JFK). Before that, he'd drawn up the majestic Gateway Arch that would loom over St. Louis. In Yorktown, it was as if he had laid the Gateway Arch on its side. The building, with three stories of glass walls, curved along the top of the hill. For visitors strolling the wide corridors decades later, the combination of the structure's rough stone and the broad vistas of rolling hills still delivered just the right message of wealth, vision, and permanence.

The idea for a *Jeopardy* machine, at least according to one version of the story, dates back to an autumn day in 2004. For several years, top executives at the company had been pushing researchers to come up with the next Grand Challenge. In the '90s, the challenge had been to build a computer that would beat a grand champion in chess. This produced Deep Blue. Its 1997 victory over Garry Kasparov turned into a global event and fortified IBM's reputation as a giant in cutting-edge computing. (This grew more important as consumer and Web companies, from Microsoft to Yahoo!, threatened to steal the spotlight—and the young brainpower. Google was still just a couple of grad students at Stanford.) Later, in another Grand Challenge in the first years of the new century, IBM produced Blue Gene, the world's fastest supercomputer.

What would the next challenge be? On that fall day, a senior manager at IBM Research named Charles Lickel drove north from his lab, up the Hudson, to the town of Poughkeepsie, and spent the day with a small team he managed. That evening, the group went to the Sapore Steakhouse in nearby Fishkill, where they could order venison, elk, or buffalo, or split a whopping fifty-two-ounce porterhouse steak for two. There, something strange happened. At seven o'clock, many of the diners stood up from their tables, their food untouched, and filed into the bar, which had a television set. "The dining room emptied," Lickel said. People were packed in there, three rows deep, to see whether Ken Jennings, who had won more than fifty straight matches on *Jeopardy,* would win again. He did. A half hour later, the crowd returned to their food, raving about the question-answering phenom. As Lickel noted, their steaks had to have been stone cold.

Though he hadn't watched much *Jeopardy* since he was a kid, that scene in the bar gave him an idea for the next Grand Challenge. What if an IBM computer could beat Ken Jennings? (Other accounts have it that the vision for a *Jeopardy* computer was already circulating along the corridors of the Yorktown lab. The original idea, it turns out, is tough to trace.)

In any event, Lickel pushed the idea. In the first meeting, it provoked plenty of dissent. Chess was nearly as clean and timeless as mathematics itself, a cerebral treasure handed down through the ages. *Jeopardy*, by contrast, looked questionable from the get-go. Produced by a publicly traded company, Sony, and subject to ratings and advertisers, it was in the business of making money and pleasing investors. It was Hollywood, for crying out loud. "There was a lot of doubt in the room," Lickel said. "People wanted something more obviously scientific." A second argument was perhaps more compelling: people playing *Jeopardy* would in all likelihood annihilate an IBM machine. "They all grabbed me after the meeting," Lickel recalled, "and said, 'Charles, you're going to regret this.'"

In the end, it was up to Paul Horn. A former professor of physics at the University of Chicago, Horn had headed IBM's three-thousand-person research arm since 1996. "If you think about smart machines," he later said, "Blue Gene by some measures had the raw computing power of the human brain, at least within an order of magnitude or two." Horn discussed those early days in his sun-splashed office at New York University, where he took up residence after his 2008 retirement from IBM. He had a black beard, and a tiny ponytail poked out from the back of his head.

"So here we have a machine that's as fast as your brain,

or close," he said. "But it doesn't think the way we think. So what would be an appropriate grand challenge that would have high visibility and excite people?" He didn't remember the idea coming from Lickel or hearing about the Fishkill dinner. In fact, Horn thought the idea might have come from him. In any case, he liked it—and promptly ran into resistance. "The general response was negative," he recalled. "People said, 'It can't be done. It's too much of a publicity stunt. The only reason that you're interested in it is because it's a show on TV.'" But Horn thought that building a savvy answering machine was the ideal challenge for IBM. While he maintained that he viewed the grand challenge as pure research, it also made plenty of sense.

IBM's business had undergone a radical transformation over the course of Horn's thirty-year career at the company. As late as the 1970s, IBM ruled the computer industry. It launched its first computers for business in 1952. But it was its breakthrough mainframe in 1964, Series 360, that established a single standard of computing in business, industry, and science. IBM pitched itself as a safe, if expensive, bet for companies looking to computerize. Its buttoned-down sales and consulting teams spread a compelling message around the world: "Nobody ever got fired for buying IBM." Big Blue, a name derived from the massive blue mainframes it sold, grew so big that its rivals, including Sperry, Burroughs, Honeywell, and four other companies, came to be known as the Seven Dwarfs. During this time, IBM researchers at Saarinen's edifice and at other labs around the world churned out an array of new technologies. They came up with magnetic strips for credit cards and floppy disks for computer data storage. Yet it was computers that drove the business. When Horn arrived at IBM Research in 1979, the greatest threat to IBM appeared to be a decade-long antitrust complaint brought by the U.S. Justice Department. It alleged that IBM had violated the Sherman Act by attempting to monopolize the fast-growing industry for business computers. Whether or not Big Blue had broken the law, its dominance was beyond question.

By 1982, when the Justice Department dropped the suit for lack of evidence, the computer world was shifting under Big Blue's feet. The previous year, IBM had unveiled its first personal computer, or PC. Priced at $1,500, it provided both legitimacy and a standard for the young industry. Early on, as corporate customers gobbled up PCs, it seemed as though IBM would go on to dominate this next stage of computing. But there was a crucial difference between these desktop machines and the mainframes. Nearly every component of the mainframes, including their processors and software, was made by IBM. In the lingo of the industry, the computers were vertically integrated. This was not the case with PCs. In order to get to market quickly at a low price, IBM built them from off-the-shelf technology—microprocessors from Intel and a rudimentary operating system, MS-DOS, from a Seattle startup called Microsoft. Since the PC had commodity innards, it took no time at all for newcomers, including Compaq and Taiwan's Acer, to plug them into cheaper "IBM-compatible" computers, or clones. IBM found itself slugging it out with a slew of upstarts while Intel and Microsoft ran away with the profits and grew into titans. Big Blue was in decline, falling faster than most people imagined. And in 1992, the vast industrial behemoth stunned the business world by registering a $4.97 billion loss, the largest in U.S. history at the time. In the space of a decade, a company that had been synonymous with cutting-edge technology now looked tired and wasteful, a manufacturing titan ill-suited to the Information Age. It almost went under.

A new chief executive, Louis V. Gerstner, arrived in 1993 and transformed IBM. He sold off or shuttered old manufacturing divisions and steered the company toward businesses based on information. IBM did not have to sell machinery to be a leader in technology, he said. It could focus on the intelligence to run the technology—the software—along with the know-how to put the systems to good use. That was services, including consulting, and it led IBM back to growth.

Technology, in the early '90s, was convulsing entire industries and the new World Wide Web promised even more dramatic change. IBM's customers, which included virtually every blue-chip company on the planet, were confused about how these new networks and services fit into their businesses. Did it make sense to shift design work to China or India and have teams work virtually? Should they remake customer service around the Web? They had loads of questions, and IBM decided it could sell the answers. It could even take over tech operations for some of its customers and charge for the service.

This push toward services and software continued under Gerstner's successor, Samuel J. Palmisano. Two months after Charles Lickel came back from Poughkeepsie with the idea for a *Jeopardy* computer that could play *Jeopardy*, IBM sold its PC division to Lenovo Group of China. That year IBM Global Services registered $40 billion in sales, more than the $31 billion in hardware sales and a much larger share of profits. (By 2009, services would grow to $55 billion, nearly 60 percent of the company's revenue. And the consultants working in the division sold lots of IBM software, which registered

$21 billion in sales.) Naturally, a *Jeopardy* computer would run on IBM hardware. But the heart of the system, like IBM itself, would be the software created to answer difficult questions.

A *Jeopardy* machine would also respond to another change in technology: the move toward human language. For most of the first half-century of the computer age, machines specialized in orderly rows of numbers and words. If the buyers in a database were listed in one column, the products in another, and the prices in a third, everything was clear: Computers could run the numbers in a flash. But if one of the customers showed up as "Don" in one transaction and "Donny" in another, the computer viewed them as two people: The two names represented different strings of ones and zeros, and therefore Don ≠ Donny. Computers had no sense of language, much less nicknames. In that way, they were clueless. The world, and all of its complexity, had to be simplified, structured and spoon-fed to these machines.

But consider what hundreds of millions of ordinary people were using computers for by 2004. They were e-mailing and chatting. Some were signing up for new social networks. (Facebook launched in February of that year.) Online humanity was creating mountains of a messy type of digital data: human language. Billions of words were rocketing through networks and piling up in data centers. Those words expressed what millions of people were thinking, desiring, fearing, and scheming. The potential customers of IBM's clients were out there spilling their lives. Entire industries grew by understanding what people were saying and predicting what they might want to do, where they might want to go, and what they were eager to buy. Google was already mining and indexing words on the Web, using them to build a media and advertising empire. Only months earlier, Google had debuted as a publicly traded company, and the new stock was sky-rocketing.

IBM wasn't about to mix it up with Google in the commercial Web. But Big Blue needed state-of-the-art tools to provide its corporate customers with the fastest and most insightful read of the words cascading through their networks. To keep a grip on its gold-plated consulting business, IBM required the very smartest, language-savvy technology—and it needed its customers to know and trust that it had it. It was central to IBM's brand.

So in mid-2005 Horn took up the challenge with a number of his top researchers, including Ferrucci. A twelve-year veteran at the company, Ferrucci managed a handful of research teams, including the five people who were teaching machines to answer simple questions in English. Their discipline was called question-answering. Ferrucci knew the challenges all too well. The machines stumbled in understanding English and appeared to plateau, in competitions sponsored by the U.S. government, at a success rate of about 35 percent.

Ferrucci wasn't a big *Jeopardy* fan, but he was familiar with it enough to appreciate the obstacles involved. *Jeopardy* tested a combination of knowledge, speed, and accuracy, along with game strategy. The show featured three contestants, each with a buzzer. In the course of about twenty minutes, they raced to respond to sixty clues representing a combined value of $54,000. Each one—and this was a *Jeopardy* quirk—was in fact an answer, some far more complex than others. The contestant had to provide the missing question. For example, in an unusual Tournament of Champions game that aired in November 1994, contestants were presented with this $500 clue* under the category Furniture: "French term for a what-not, a stand of tiered shelves with slender supports used to display curios." The host, Alex Trebek, read the clue from the big game board. The moment he finished, a panel around the question lit up setting off the race to buzz. On average, contestants had about four seconds to read and consider the clue before buzzing. The first to buzz was, in effect, placing a bet. The right response—"What is an étagère?"—was worth $500 and gave the contestant the chance to pick again. ("Let's try European Capitals for $200.") A botched response wiped the same amount from a contestant's score and gave the other two a chance to try. (In this example, no one dared to buzz. Such a clue, uncommon in *Jeopardy*, is known as a "triple-stumper.")

To compete in *Jeopardy*, a machine not only would need to come up with the answer, posed as a question, within four seconds, but it would also have to gauge its confidence in its response. It would have to know what it knew. "Humans know what they know like *that*," Ferrucci said later, snapping his fingers. Replicating such confidence in a computer would be tricky. What's more, the computer would have to calculate the risk according to where it stood in the game. If it was far ahead and had only middling confidence on "étagère," it might make more sense not to buzz. In addition to piling up knowledge, a computer would have to learn to play the game.

Complicating the game strategy were four wild cards. Three of the game's sixty hidden clues were so-called Daily Doubles. In that 1994 game, a contestant named Rachael

---

* In *Jeopardy*, the answers on the board are called "clues," and the players' questions—what most viewers perceive as answers—are "responses."

Schwartz, an attorney from Bedminster, New Jersey, asked for the $400 clue in the Furniture category. Up popped a Daily Double giving her the chance to bet some or all of her money on a furniture-related clue she had yet to see. She wagered $500, a third of her winnings, and was faced with this clue: "This store fixture began in 15th century Europe as a table whose top was marked for measuring." She missed it, guessing, "What is a cutting table?," and lost $500. ("What is a counter?" was the correct response.) It was early in the game and didn't have much impact. The three players were all around the $1,000 mark. But later in a game, Ferrucci saw, Daily Doubles gave contestants the means to storm back from far behind. A computer playing the game would require a clever game program to calibrate its bets.

The biggest of the wild cards was Final Jeopardy, the last clue of the game. As in Daily Doubles, contestants could bet all or part of their winnings on a single category. But all three contestants participated—as long as they had positive earnings. Often the game boiled down to betting strategies in Final Jeopardy. Take that 1994 contest, in which the betting took a strange turn. Going into Final Jeopardy, Rachael Schwartz led Kurt Bray, a scientist from Oceanside, California, by a slim margin, $9,200 to $8,600. The category was Historic Names. To lock down a win, she had to assume he would bet everything, reaching $17,200. A bet of $8,001 would give her one dollar more, provided she got it right. But if they both bet big and missed, they might fall to the third-place contestant, Brian Moore, a Ph.D. candidate from Pearland, Texas. In the minute or so that they took to place their bets, the two leaders had to map out the probabilities of a handful of different scenarios. They wrote down their dollar numbers and waited for the clue: "Though he spent most of his life in Europe, he was governor of the Bahamas for most of World War II."

The second-place player, Bray, was the only one to get it right: "Who was Edward VIII?" Yet he had bet only $500. It was a strange number. It placed him $100 behind the leader, not ahead of her. But the bet kept him beyond the reach of the third-place player. Most players bet at least something on a clue. If Schwartz had wagered and missed, he would win. Indeed, Schwartz missed the clue. She didn't even bother guessing. But she had bet nothing, leaving herself $100 ahead and winning the game.

The betting in Final Jeopardy, Ferrucci saw, might actually play to the strength of a computer. A machine could analyze betting patterns over thousands of games. It could crunch the probabilities and devise optimized strategies in a fraction of a second. "Computers are good at that kind of math," he said.

It was the rest of *Jeopardy* that appeared daunting. The game featured complex questions and a wide use of puns posing trouble for literal-minded computers. Then there was *Jeopardy*'s nearly boundless domain. Smaller and more specific subject areas were easier for computers, because they offered a more manageable set of facts and relationships to master. They provided context. A word like "leak," for example, had a specific meaning in deep-sea drilling, another in heart surgery, and a third in corporate press relations. A know-it-all computer would have to recognize different contexts to keep the meanings clear. And *Jeopardy*'s clues took the concept of a broad domain to a near-ludicrous extreme. The game had an entire category on Famous Understudies. Another was on the oft-forgotten president Rutherford B. Hayes. Worse, from a computer architect's point of view, the game demanded answers within seconds—and penalized players for getting them wrong. A *Jeopardy* machine, just like the humans on the show, would have to store all of its knowledge in its internal memory. (The challenge, IBM figured, wouldn't be nearly as impressive if a bionic player had access to unlimited information on the Web. What's more, *Jeopardy* would be unlikely to accept a Web-surfing contestant, since others didn't have the same privilege.) Beating humans in *Jeopardy*, it seemed, was more than a stretch goal. It appeared impossible and spelled potential disaster for researchers. To embarrass the company on national television—or, more likely, to flame out before even getting there—was no way to manage a career.

Ferrucci's pessimism was also grounded in experience. In annual government competitions, known as TRec (Text Retrieval Conference), his question-answering (Q-A) team developed a system called Piquant. It struggled far below *Jeopardy* levels with a much easier test. In TRec, the competing teams were each given a relatively small "corpus" of about one million documents. They then had to train the machines to answer questions based on the material. (In one version from 2004, several of the questions had to do with Tom Cruise and his ex-wife.)

In answering these questions, the computer, for all its processing power and memory, resembled nothing so much as a student with serious brain damage. An apparently simple question could turn it into knots. In 2005, it was asked: "What is Francis Scott Key best known for?" The first job was to determine which of those words represented the subject of the question, the "entity," and whether that might be a person, a state, or perhaps an animal or a machine. Each one had different characteristics. "Francis" and "Scott" looked like names. But "Key"? That could be a metal tool to open doors or a mental breakthrough to solve problems. In its hunt, the computer might even spend a millisecond or two puzzling over Key lime pies. Clearing up these doubts might require a visit to the system's "disambiguation" unit, where the answering program consulted a dictionary or looked for contextual clues in the surrounding words. Could "Key" be something the ingenious Francis Scott invented, collected, planted, or stole? Could he have baked it? Probably not. The structure of the question, with no direct object, made it look like the third name of a person. The capital K on Key strengthened that case.

A person confronting that question either knew or did not know that Francis Scott Key wrote the U.S. national anthem, "The Star-Spangled Banner." But he or she wasted no time searching for the subject and object in the sentence or wondering if it was a last name, a metal tool, or a tangy South Florida dessert.

For the machine, things only got worse. The question lacked a verb, which could disorient the computer. If the question were, "What did Francis Scott Key *write*?" the machine could likely find a passage of text with Key writing something, and that something would point to the answer. The only pointer here—"is known for"—was maddeningly vague. Assuming the computer had access to the Internet (a luxury it wouldn't have on the show), it headed off with nothing but the name. In Wikipedia, it might learn that Key was "an American lawyer, author and amateur poet, from Georgetown, who wrote the words to the United States national anthem, 'The Star-Spangled Banner.'" For humans, the answer was right there. But the computer, with no verb to guide it, might answer that Key was known as an amateur poet or a lawyer from Georgetown. In the TRec competitions, IBM's Piquant botched two out of every three questions.

All too often, the system failed to understand the question or to put it in the right context. For this, a growing school of Artificial Intelligence argued, systems needed to spend more time in the computer equivalent of infancy, mastering the concepts that humans take for granted: time, space, and the basic laws of cause and effect.

Toddlerhood is a tribulation for computers, because it represents knowledge that is tied to the human experience: the body and the senses. While crawling, we learn about space and physical objects, and we get a sense of time. The toddler reaches for the jar on the table. Moments later pieces of it lie scattered on the floor. What happened between those two states? It fell. Such lessons establish notions of before and after, cause and effect, and the nature of gravity. These experiences, most of them accompanied by a steady stream of human language, set the foundation for practically everything we learn. "You crawl around and bump into things," said David Gunning, a senior manager at Vulcan Inc., an AI incubator in Seattle. "That's basic research." It isn't just jars that fall, the toddler notices. Practically everything does. (Certain balloons are exceptions, which seem magical.) The child turns these observations into theory. Unlike computers, humans generalize.

Even the metaphors in our language lead back to the tumbles and accidents seared into our consciousness in our early years. We "fall" for a sales pitch or "fall" in love, and we cringe at hearing "sharp" words or "stinging" rebukes. We process such expressions on such a basic level that they seem closer to feeling than thought (though for humans, unlike computers, the two are intertwined). Over the course of centuries, these metaphors infused language and, consequently, were fundamental to understanding *Jeopardy* clues. Yet to a machine with no body or experience in the physical world, each one was a puzzle.

In some Artificial Intelligence labs, scientists were attempting to transmit these elementary experiences to computers. Sajit Rao, a professor at MIT, was introducing computers equipped with vision to rumpus-room learning, showing them objects moving, falling, obstructing paths, and piling on top of one another. The goal was to establish a conceptual understanding so that eventually computers could draw conclusions from visual observations. What would happen, for example, when vehicles blocked a road?

Several years later, the U.S. Defense Department's Advanced Research Projects Agency (DARPA) would fund Rao's research for a program called Mind's Eye. The idea was to teach machines not only to recognize objects but to be able to reason about what they were doing, where they might have come from. This work, they hoped, would lead to smart surveillance cameras, which would mean that computers could replace humans in the tedious and exhausting task of monitoring a spot—what the Pentagon calls "persistent stare." Instead of simply recording movements, these systems would interpret them. If a man in Afghanistan went into a building carrying a package and emerged without it, the system would conclude that he had left it there. If he walked toward another person with a suitcase in his hand, it would predict that he was going to give it to him. A seeing and thinking machine that could generate hypotheses based on observations might zero in on potential roadside bombs or rooftop snipers. This type of intelligence, according to DARPA, would extend computer surveillance from objects to actions—from nouns to verbs.

This skill required the computer to understand relationships—precisely the stumbling block of IBM's Piquant as it struggled with questions in the TRec competition. But potential breakthroughs such as Mind's Eye were still in the infant stage of research and wouldn't be ready for years—certainly not in time to give a *Jeopardy* machine a dose of human smarts. What's more, Ferrucci was busy managing another big software project. So after consulting his team and assembling the discouraging evidence, he broke the news to a disappointed Paul Horn. His team would not pursue the *Jeopardy* challenge. It was just too hard to guarantee results on a schedule.

Free of that distraction, the Q-A team returned to its work, preparing Piquant for the next TRec competition. As it turned out, though, Ferrucci had won them only a respite, and a short one at that. Months later, in the summer of 2006, Horn returned with exactly the same question: How about *Jeopardy*?

Reluctantly, Ferrucci and his small Q-A team gathered in a small room at the Hawthorne research center, a ten-minute drive south from Yorktown. (It was a far less elegant structure, a cuboid of black glass in an office park. But unlike Yorktown, where the public spaces were bathed in natural light and the offices windowless, Hawthorne's offices did have views, mostly of parking lots.) The discussion followed the familiar, depressing lines: the team's travails in the TRec competitions, the insanely broad domain of *Jeopardy*, and the difficulty of coming up with answers and a betting strategy in three to five seconds. TRec had no time limit at all, and the computer often churned away for minutes trying to answer a single question.

While the team talked, Ferrucci sat at the back of the room, uncharacteristically quiet. He had a laptop open and was typing away. He was looking up *Jeopardy* clues online and then searching for answers on Google. The answers certainly didn't pop up. But in many cases, the search engine led to the right neighborhood. He started thinking about the technologies needed to refine Google's vague pointer to a precise answer. It would require much of the tech muscle of IBM. He'd have to bring in top natural-language researchers and experts in machine learning. To speed up the answering process, he'd need to spread out the computing to hundreds or even thousands of machines. This would require a crack hardware unit. His team would also need to educate the machine in strategy. Ferrucci had a few colleagues who focused on game theory. Several of them were training computers to play the Japanese game Go (whose computational complexity made chess look like Tic-Tac-Toe). Putting together all the pieces of this electronic brain would require a large multidisciplinary team and a huge investment—and even then they might fail. But the prospect of success, however remote, was tantalizing. Ferrucci looked up from his computer and said "Hey, I think we can do this."

At the dawn of Artificial Intelligence (AI), a half century ago, scientists predicted that computers would soon be speaking and answering questions fluently. A pioneer in the field, Herbert Simon, predicted in 1965 that "machines w[ould] be capable, within twenty years, of doing any work a man can do." These were the glory days of AI, a period of boundless vision and bounteous funding. Machines, it seemed, would soon master language, recognize faces, and maneuver, as robots, in factories, hospitals, and homes. In short, computer scientists would create a superendowed class of electronic servants. This led, of course, to failed promises, to such a point that Artificial Intelligence became a term of derision. Bold projects to build bionic experts and conversational computers lost their sponsors. A long AI winter ensued, lasting through much of the '80s and '90s.

What went wrong? In retrospect, it seems almost inconceivable that leading scientists, including Nobel laureates like Simon, believed it would be so easy. They certainly appreciated the complexity of the human brain. But they also realized that a lot of that complexity was tied up in dreams, memories, guilt, regrets, faith, desires, along with the controls to maintain the physical body. Machines wouldn't have to bother with those details. All they needed was to understand the elements of the world and how they were related to one another. Machines trained in the particulars of sick people; ambulances and hospitals, for example, could conceivably devote their analytical skills to optimizing emergency services. Yet teaching the machines proved extraordinarily difficult. One of the biggest challenges was to anticipate the responses of humans. The machines weren't up to it. And they had serious trouble with even the most basic forms of perception, such as seeing. For example, researchers struggled to teach machines to perceive the edges of things in the physical world. As it turned out, it required experience and knowledge and advanced powers of pattern recognition just to look through a window and understand that the oak tree in the yard was a separate entity. It was not connected to the shed on the other side of it or a pattern on the glass or the wallpaper surrounding the window.

The biggest obstacle, though, was language. In the early days, it looked beguilingly easy. It was just a matter of programming the machine with vocabulary and linking it all together with a few thousand rules—the kind you'd find in a grammar book. If the machine still underperformed? Well, just give it more vocabulary, more rules.

Once the electronic brain mastered language, it was simply a question of teaching it about the world. Asia's over there. This is the United States. We have a democracy. That's the Pacific Ocean between the two. It's big, and wet. If researchers kept adding facts, millions of them, and defining their relationships, by the end of the grant cycle they might have a talking, thinking machine that "knew" what humans did.

Language, of course, turns out to be far more complicated. Jaime Carbonell, a top researcher at Carnegie Mellon University, has been teaching language to machines for decades. The way he describes it, our minds are swimming with cultural and historical allusions, accumulated over millennia, along with a complex scheme of who's who. Words, when spoken or read, vary wildly according to context. (Just imagine if the cops in New York raced off to Citi Field, sirens wailing, every time someone was heard saying, "The Mets are getting killed!")

Carbonell, sitting in his Pittsburgh office, gave another example. He issued a statement: "I want a juicy hamburger." What does it mean? Well, if a child says it to his mother, it's a request or a plea. If a general says it to a corporal, it's a tacit command. And if a prisoner says it to a cellmate, it might be nothing more than a wish. Scientists, of course, could attempt to teach a computer those variables as rules. But new layers of complexity pop up. Is the general a vegan or speaking sarcastically? Or maybe "hamburger" means something entirely different in prison lingo?

This flexibility isn't a weakness of language but a strength. Humans need words to be inexact; if they were too precise, each person would have a unique vocabulary of several billion words, all of them unintelligible to everyone else. You might have a unique word for the sip of coffee you just took at 7:59 A.M., which was flavored with the anxiety about the traffic in the Lincoln Tunnel or along Paris's Périphérique. (That single word would be as useless to you as to everyone else. A word has to be used at least twice to have any purpose.)

Each word is a lingua franca, a fragment of a clumsy common language. Imagine a man saying a simple sentence to a friend: "I'm weary." He's thinking about something, but what is it? Has he carried a load a long way in the sun? Does he have a sick child or financial troubles? His friend certainly has different ideas, based on his own experience, about what "weary" means. In addition to the various contexts, it might send other signals. Maybe where he comes from, the word has a slightly rarefied feel, and he's wondering whether his friend is trumpeting his sophistication. Neither one knows exactly what the other is thinking. But that single word, "weary," extends an itsy bridge between them.

Now, with that bridge in place, the word shared, they dig deeper to see if they can agree on its meaning. They study each other's expression and tone of voice. As Carbonell noted, context is crucial. Someone who has won the Boston Marathon might be contentedly weary. Another, in a divorce hearing, is anything but. One person may slack his jaw in an exaggerated way, as if to say "Know what I mean?" In this tiny negotiation, far beyond the range and capabilities of machines, two people can bridge the gap between the formal definition of a word and what they really want to say.

It's hard to nail down the exact end of AI winter. A certain thaw set in when IBM's computer Deep Blue bested Garry

Kasparov in their epic 1997 showdown. Until that match, human intelligence, with its blend of historical knowledge, pattern recognition, and the ability to understand and anticipate the behavior of the person across the board, ruled the game. Human grandmasters pondered a rich set of knowledge, jewels that had been handed down through the decades—from Bobby Fischer's use of the Sozin Variation in his 1972 match with Boris Spassky to the history of the Queen's Gambit Declined. Flipping through scenarios at about three per second—a glacial pace for a computing machine—these grandmasters looked for a flash of inspiration, an insight, the hallmark of human intelligence.

Equally important, chess players tried to read the minds of their foes. This is a human specialty, a mark of our intelligence. Cognitive scientists refer to it as "theory of mind"; children develop it at about age four. It's what enables us to imagine what someone else is experiencing and to build large and convoluted structures based on such analysis. "I wonder what he was thinking I knew when I told him . . ." Most fiction, from Henry James to Elmore Leonard, revolves around this very human analysis, something other species—and computers—cannot even approach. (It's also why humans make such expert liars.)

Unlike previous AI visions, in which a computer would "think" more or less the way we do, Deep Blue set off on a different course. It played on the strengths of a supercomputer: a fabulous memory and extraordinary calculating speed. Statistical approaches to machine intelligence had been around since the dawn of AI, but the numbers mavens had never witnessed anything approaching this level of computing power and speed. Deep Blue didn't try to read Garry Kasparov's mind, and it certainly didn't count on flashes of inspiration.

Instead, it raced through a century of grandmaster games, analyzing similar moves and situations. It then constructed the most probable scenarios for each possible move. It analyzed two hundred million moves per second (nearly seventy million for each one the humans considered). A similar approach for a computer writing poetry would be to scrutinize the patterns and vocabulary of every poem ever written before choosing each word.

Forget inspiration, creativity, or blinding insight. Deep Blue crunched data and won its match by juggling statistics, testing thousands of scenarios, and calculating the odds. Its intelligence was alien to human beings—if it could be considered intelligence at all. IBM at the time described the machine as "less intelligent than the stupidest person." In fact, the company stressed that Deep Blue did not represent AI, since it didn't mimic human thinking. But the Deep Blue team made good on a decades-old promise. They taught a machine to win a game that was considered uniquely human. In this, they passed a chess version of the so-called Turing test, an intelligence exam for machines devised by Alan Turing, a pioneer in the field. If a human judge, Turing wrote, were to communicate with both a smart machine and another human, and that judge could not tell one from the other, the machine passed the test. In the limited realm of chess, Deep Blue aced the Turing test—even without engaging in what most of us would recognize as thought.

But knowledge? That was another challenge altogether. Chess was esoteric. Only a handful of specially endowed people had mastered the game. Yet all of us played the knowledge game. By advancing from chess to *Jeopardy*, IBM was shifting the focus from a remote island off the coast straight to our cognitive mainland. Here, the computer would grapple with far more than game theory and math. It would be competing in a field utterly defined by human intelligence. The competitors in *Jeopardy,* as well as the other humans writing the clues, would feast on knowledge tied to experiences and sensations, sights and tastes. The machine, by contrast, would be blind and deaf, with no body, no experience, no life. Its only memories—if you could call them that—would be millions of lists and documents encoded in ones and zeros. And the entire game would be played in endlessly complex and nuanced language—a cinch for humans, a tribulation for machines.

Picture one of those cartoons in which a land animal, perhaps a coyote, runs off a cliff and continues to run so fast in midair that it manages to fly (at least for a while). Now imagine that animal not only surviving but flying upward and competing with birds. That would be the challenge facing an IBM machine. It would have to use its native strengths in speed and computation to thrive in an utterly foreign setting. Strictly speaking, the machine would be engaged in a knowledge game without "knowing" a thing.

Still, Ferrucci believed his team had a fighting chance, though he wasn't quite ready to commit. He code-named the project Blue J—Blue for Big Blue, J for *Jeopardy*—and right before the holidays, in late 2006, he asked Horn to give him six months to see if it was possible.

2

And Representing the Humans . . .

ON A LATE SUMMER day in 2004, a twenty-nine-year-old software engineer named Ken Jennings placed a mammoth $12,000 bet on a Daily Double in *Jeopardy*. The category was Literary Pairs. Jennings, who by that point had won a record fifty straight games, was initially flummoxed by the clue: "The film title 'Eternal Sunshine of the Spotless Mind' comes from a poem about these ill-fated medieval lovers." As the seconds passed, Jennings flipped through every literary medieval couple he could conjure up—Romeo and Juliet, Dante and Beatrice, Petrarch and Laura—but he found reasons to disqualify each one. Time was running out. A difference of $24,000 was at stake, enough for a new car. Jennings quickly reviewed the clue. On his second reading, something about the wording suggested to him that the medieval lovers were historical figures, not literary characters or their creators. He said he couldn't put his finger on it, but it had "a flavor of history." At that point, the names of the French philosopher Peter Abelard and his student and lover, Heloise, popped into Jennings's mind. It was the answer. He just knew it. It was their correspondence, the hundreds of letters they exchanged after their tragic separation, that qualified them as a literary pair. He pronounced the names as time ran out, pocketed the $12,000, and moved on to the next clue.

In answering that single clue, Jennings displayed several peerless qualities of the human mind, ones that IBM's computer engineers would be hard-pressed to instill in a machine. First, he immediately understood the complex clue. Unlike even the most sophisticated computers, he was a master of human language. Far beyond basic comprehension, he picked up nuance in the wording so very subtle that even he failed to decode it. Yet it pushed him toward the answer. Once Abelard and Heloise surfaced, more human magic kicked in: He knew he was right. While a *Jeopardy* computer would no doubt weigh thousands, even millions, of possibilities, humans could look at a mere handful and often pick the right one with utter confidence. Humans just know things. And good *Jeopardy* players often sense that they'll find the answer, even before it comes to mind. "It's an odd feeling," Jennings wrote in his 2005 memoir, *Brainiac*. "The answer's not on the tip of your tongue yet, but a light flashes in the recesses of your brain. A connection has been made, and you find your thumb pressing the buzzer while the brain races to catch up."

Perhaps the greatest advantage humans would enjoy over a *Jeopardy* machine was kinship with the fellow humans who had written the clues. With each clue, participants attempt to read the mind of the writers. What response could they be looking for? In an easy $200 category, would the writers expect players to recognize a Caribbean nation as small as Saint Lucia? With that offhand reference to "candid," could they be pointing toward Voltaire's *Candide*? Would they ever stack the European Capitals category with two clues featuring Dublin? When playing *Jeopardy,* Jennings said, "You're not just parsing the question, you're getting into the head of the writer." In this psychological aspect of the game, a computer would be out of its league.

Computers, of course, can rummage through mountains of data millions of times faster than humans. But humans compensate with mental shortcuts, many of them honed over millions of years of evolution. Instead of plowing through copious evidence, humans instinctively read signals and draw quick conclusions, whether they involve trusting a stranger or deciding where to pitch a tent. "Mortals cannot know the world, but must rely on uncertain inferences, on bets rather than demonstrable proof," wrote the German psychologist Gert Gigerenzer. In recent decades, psychologists have unearthed dozens of these rules, known as heuristics. Many of them would guide humans in a *Jeopardy* match against a much faster computer.

The most elementary heuristic is based on favoring the first answer to pop into the brain. That one automatically starts in the front of the line; it is more trusted simply by virtue of arriving early. Which ideas pop in first? Following another heuristic, they're often the answers contestants are most familiar with. Given a choice between a well-known place or person or an obscure one, studies show that people opt for what they know. "If you ask people, 'Which of these two cities has a larger population,' they'll almost always choose the more familiar one," said Richard Carlson, a professor of cognitive psychology at Penn State. Usually this works. If a *Jeopardy* player has to name the most populous cities in a certain country, the most famous ones—London, Tokyo, Berlin, New York—often fit the bill. This approach can lead to bloopers, of course. But it happens less often in *Jeopardy* than in the outside world. Why? Again, the writers, being human, work from the same rules of thumb, and they're eager to connect with contestants and with the nine million people watching on TV. They want the contestants to succeed and to look smart, and they want people at home to feel smart, too. That's critical to *Jeopardy*'s popularity. "You can't forget that it's a TV show," said Roger Craig, a six-time *Jeopardy* champion. "They're writing for the person in the living room." And that viewer, like Ken Jennings—and unlike a computer—races along well-worn mental paths to answer questions. These paths are marked with signs and signals that call out to the human brain and help it navigate.

A century ago, the psychologist William James divided human thought into two types, associative and true reasoning. For James, associative thinking worked from historical patterns and rules in the mind. True reasoning, which was necessary for unprecedented problems, demanded deeper analysis. This came to be known as the "dual process" theory. Late in the twentieth century, Daniel Kahneman of Princeton redefined these cognitive processes as System 1 and System 2. The intuitive System 1 appeared to represent a primitive part of the mind, perhaps dating from before the cognitive leap undertaken by our tool-making Cro-Magnon ancestors forty thousand years ago. Its embedded rules, with their biases toward the familiar, steered people toward their most basic goals: survival and reproduction. System 2, which appeared to arrive later, involved conscious and deliberate analysis and was far slower. When it came to intelligence, all humans were more or less on an equal footing in the ancient and intuitive System 1. The rules were easy, and whether they made sense or not, everyone knew them. It was in the slower realm of reasoning, System 2, that intelligent people distinguished themselves from the crowd.

Still, great *Jeopardy* players like Ken Jennings cannot afford to ignore the signals coming from the caveman quarters of their minds. They need speed, and the easy answers pouring in through System I are often correct. But they have to know when to distrust this reflexive thought, when to pursue a longer and more analytical route. In the same game in which Jennings tracked down Abelard and Heloise, this clue popped up in the Tricky Questions category: "Total number of each animal that Moses took on the ark with him during the great flood." Jennings lost the buzz to Matt Kleinmaier, a medical student from Chicago, who answered, "What is two?" It was wrong. Jennings, aware that it was supposed to be tricky, noticed that it asked for "each animal" instead of "each species." He buzzed for a second chance at the clue and answered, "What is one?" That was wrong, too. The correct answer, which no one came up with, was "What is zero?"

Jennings and Kleinmaier had fallen for a trick. Each had focused on the gist of the clue—the number of animals boarding the biblical ark—while ignoring one detail: The ark builder was Noah, not Moses. This clue actually came from a decades-old psychological experiment, one that has given a name—the Moses Illusion—to the careless thinking that most humans employ.

It's easy enough to understand. The brain groups information into clusters. (Unlike computers, it doesn't move packets of encoded data this way and that. The data stay put and link up through neural connections.) People tend to notice when one piece of information doesn't jibe with its expected group. It's an anomaly. But Noah and Moses cohabit numerous clusters. Thematically they are both in the Bible, visually, both wear beards. Phonetically, their names almost rhyme. A question about Ezekiel herding animals into the ark might not pass so smoothly. According to a study headed by Lynn Reder, a psychologist at Carnegie Mellon, the Moses Illusion illustrates a facet of human intelligence, one vital for *Jeopardy*.

Most of what humans experience as perception is actually furnished by the memory. This is because the conscious brain can only process a trickle of data. Psychologists agree that only one to four "items," either thoughts or sensations, can be held in mind, immediately available to consciousness, at the same time. Some have tried to quantify these constraints. According to the work of Manfred Zimmerman of Germany's Heidelberg University, only a woeful fifty bits of information per second make their way into the conscious brain, while an estimated eleven million bits of data flow from the senses every second. Many psychologists object to these attempts to measure thoughts and perceptions as digital bits. But however they're measured, the stark limits of the mind are clear. It's as if each person's senses generated enough data to run a 3D Omnimax movie with Dolby sound—only to funnel it through an antediluvian modem, one better suited to Morse code. So how do humans re-create the Omnimax experience? They focus on the items that appear most relevant and round them out with stored memories, what psychologists call "schemas."

In the Moses example, people concentrate on the question about animals. The biblical details, which appear to fit into their expected clusters, are ignored. It's only when a wrong name intrudes from outside the expected orbit that alarms go off. In one experiment at Carnegie Mellon, when researchers substituted a former U.S. president for Moses, people noticed right away. Nixon had nothing to do with the ark, they said.

Even after falling victim to the Moses Illusion, Jennings found no fault in his own thinking. "The brain's doing the right thing!" he said. "It's focusing on the right part of the question: How many animals did the biblical figure take onto the ark?" That, he said, is how the brain *should* work. "It's just that the question writer has found a way to work against you." Those sorts of tricks, he added, are uncommon on *Jeopardy*.

Strangely enough, the cerebral carelessness that leads to the Moses Illusion also serves a useful function for human thought. Filtering out details not only eliminates time-consuming busy work. It also allows people to overlook many variations and to generalize. This is important. If they focus too much on small changes, they might think, for example, that each time a friend gets a haircut or a suntan, she's a different person. Instead, the brain settles on the gist of the person and is ready to look past some details—or, in many cases, to ignore them. This can be embarrassing. (Sometimes it *is* a different person.) Still, by skipping over details, the brain is carrying out a process that is central to human intelligence and one that confounds computers. It's thinking more broadly and focusing on concepts.

The *Jeopardy* studio sits on the sun-drenched Sony lot in Culver City. Seven miles south of Hollywood's Sunset and Vine, this was a suburban hinterland when Metro-Goldwyn-Mayer (MGM) started making movies there in 1915. In later decades it turned out such classics as *The Wizard of Oz* and *Ben Hur*—all of them introduced by the iconic roaring lion. Following years of mergers and acquisitions, the lot became the property of a Japanese industrial giant—a development that likely would have shocked Samuel Goldwyn. Sony later gobbled up Columbia Studios, which had belonged to Coca-Cola for a few years in the eighties. On the Sony lot, the MGM lion gave way to Lady Liberty holding her torch. In the summer of 2007, as IBM considered a *Jeopardy* project, tourists on the Sony lot were filing past the sets of *Spiderman II* and Will Smith's *Happyness*. Others with free passes lined up for *Jeopardy*. If they made their way past the fake Main Street, with its cinema, souvenir shop, and café, they would come across a low-slung office building named for Robert Young, the actor who played the homespun 1970s doctor Marcus Welby, M.D.

This is where Harry Friedman worked. Friedman, then in his late fifties, was the executive producer of both *Wheel of Fortune* and *Jeopardy,* the top- and second-ranked game shows in America. *Wheel,* as it was known, relied on the chance of a spinning wheel and required only the most rudimentary knowledge of common phrases and titles. Its host was a former TV weatherman named Pat Sajak, who had been accompanied since 1983 by the lovely Vanna White. She had showcased more than four thousand dresses through the years while turning the letters on the big board and leading the clapping while the roulette wheel spun. For some *Jeopardy* fans, even mentioning the two games in the same breath was an outrage. It would be like card players comparing the endlessly complex game of Bridge to Go Fish. Nevertheless, *Wheel* attracted some eleven million viewers every weeknight evening, and about nine million tuned in to *Jeopardy*. Harry Friedman's job, while touching on the world of knowledge and facts, was to keep those millions of people watching his two hit shows. In a media world exploding with new choices, it was a challenge.

In movie studios on this Sony-Columbia lot, men with the bookish mien of Harry Friedman are cast as professors, dentists, and accountants. His hair, which recedes toward the back of his head, is still dark, and matches the rims of his glasses. His love for television dates back to his childhood.

His father ran one of the first TV dealerships in Omaha, and the family had the first set in the neighborhood, a 1950 Emerson with a rounded thirteen-inch screen. Friedman's goal as a youngster was to write for TV. While he was in college, he pursued writing, working part-time as a sports and general assignment reporter for the *Lincoln Star*. After graduating, in 1971, he traveled to Hollywood. He eventually landed a part-time job at *Hollywood Squares,* a popular daytime game show, where he wrote for $5 a joke.

Friedman climbed the ladder at *Hollywood Squares,* eventually producing the show. He also wrote stand-up acts for comedians and entertainers, people like Marty Allen and Johnny Carson's old trumpet-playing bandleader, Doc Severinsen. He got his big break in 1994, when he was offered the top job at *Wheel of Fortune.* The show, a sensation in the 1980s, was stagnating. Friedman soon saw that antiquated technology had slowed the game to a crawl. The spectators, hosts, and audience had to sit and wait for ten or fifteen minutes between each round while workers installed the next phrase or jingle with big cardboard letters. Friedman ordered a shift to electronic letters. The game speeded up. Ratings improved.

Two years later, he was offered the top job at *Jeopardy.* The game, which today radiates such wholesomeness, emerged from the quiz show scandals of the 1950s. "That's where we came from. That's our history," Friedman said. Back then, millions tuned their new TV sets to programs that featured intellectual brilliance. Among the most popular was *Twenty-One,* where a brainy young college professor named Charles Van Doren appeared to be all but omniscient. The ratings soared as Van Doren summoned answers. Often they came instantly. Other times he appeared to dig into the dusky caverns of his memory, surfacing with the answer only after a torturous and suspenseful mental hunt. Van Doren seemed to epitomize brilliance. He was a phenomenon, a national star. This was the kind of brainpower the United States would be needing—in technology, diplomacy, and education—to prevail over the Soviet Union in the Cold War. Knowledge was sexy. And when it turned out that the producers were feeding Van Doren the answers, a national scandal erupted. It led to congressional hearings, a condemnation by President Eisenhower—"a terrible thing to do to the American people"—and stricter regulations covering the industry. For a few years, quiz shows all but disappeared.

In 1963, Merv Griffin, the talk show host and entrepreneur, was wondering how to resurrect the format. According to a corporate history book, he was in an airplane with his wife, Julann, when the two of them came up with an idea. If people suspect that you're feeding contestants the answers, why not devise a show that provides the answers—and forces players to come up with the questions?

It was the birth of *Jeopardy*. Griffin came up with simple, enduring rules, the sixty clues, including three hidden Daily Doubles and the tiny written exam for Final Jeopardy. To fill the thirty seconds while the players scribbled their final response on a card, Griffin wrote a catchy sixty-four-note jingle that became synonymous with the show. He hired Art Fleming, a strait-laced actor in TV commercials, as the game's host. In March 1964, *Jeopardy* was launched as a daytime show. It continued through 1975 and reappeared briefly at the end of that decade.

Griffin brought *Jeopardy* back in 1984 as a syndicated evening show hosted by a young, mustachioed Alex Trebek. A new board game, Trivial Pursuit, was a national rage, and the mood seemed right for a *Jeopardy* revival. The new game was much the same—the three-contestant format, the (painfully) contrived little chats with the host following the first commercial break, and the jingle during Final Jeopardy. It took time for the new show to catch on. In its first year, it was relegated to the wee hours in many markets, including New York. But within a few years, it settled into early evening time slots. It was eventually syndicated on 210 stations and became a ritual for millions of fact-loving viewers.

Still, when Friedman arrived at *Jeopardy* in 1997, he saw a problem. Too many of the questions still focused on academic subjects. They were the same types of history, geography, and literature clues that had captivated America four decades earlier, when Charles Van Doren paraded his faux smarts. But times had changed, and so had America's intellectual appetite. Sure, some of the most dedicated viewers still subscribed to the show's mission, to inform and educate. They wanted reminders on the river that separated cisalpine Gaul from Italy in Roman times ("What is the Rubicon?"), the last British colony on the American mainland to gain independence ("What is Belize?"), and the 1851 novel that contained "a dissertation on cetology" ("What is *Moby Dick*?").

These were the *Jeopardy* purists. They tended to be older, raised in Van Doren's heyday. But their ranks were shrinking as other types of information were exploding on the brand-new World Wide Web. As Friedman put it: "Anything that veered off the academic foundation was deemed to be pop culture. And to purists, that was heresy." But he feared that *Jeopardy* would lose relevance if it relied on academic clues in an age of much broader information.

So he leavened the mix, bringing in more of the topics that consumed people on coffee breaks, from sports to soap opera. If you remembered the person who conspired in 1994 to "whack Nancy Kerrigan's knee" ("Who is Tonya Harding?"), you probably didn't learn about her while reading *Bartlett's Quotations* or brushing up on the battle of Gettysburg. Sometimes Friedman blended the popular and the scholarly. During the 1999 season, one category was called Readings from Homer. It featured clues about the other Homer, author of the *Odyssey* and the *Iliad*, read by Dan Castelleta, the voice of the lovable dunce of TV's *The Simpsons*. The clues were written in the dumbed-down style of the modern Homer: "Hero speaking here: 'Nine days I drifted on the teeming sea . . . upon the tenth we came to the coastline of the lotus eaters. . . . Mmmm, lotus!'" ("Who is Odysseus?")

From the perspective of a *Jeopardy* computer, it's worth noting that Friedman's adjustments to the *Jeopardy* canon made the game harder. Instead of mastering a set of formal knowledge, the computer would have to troll the ever-expanding universe of what modern folk carried around in their heads. This shifted the focus from what people *should* know to what they *did* know—collectively speaking—from a few shelves of reference books to the entire Internet. What's more, for a computer, the formal stuff—the factoids—tended to be far easier. Facts often appear in lists, many of them accompanied by dates. One mention of the year 1215, and any self-respecting *Jeopardy* computer could sniff out the relevant document ("What is the Magna Carta?"). But imagine a computer responding to this clue: "Here are the rules: if the soda container stops rotating & faces you, it's time to pucker up" ("What is Spin the Bottle?").

Yes, Harry Friedman turned *Jeopardy* into a tougher game for computers, and he also built it into a breeding ground for celebrity champions. Throughout its history, *Jeopardy* maintained a strict limit of five matches for returning champs.

This seemed unfair to Friedman, and he debated it with colleagues for years. The downside? "You get somebody on the show who is there forever," he said. Imagine if the person was unlikable or, worse, boring. Nonetheless, he lifted the limit in 2003. And the following year—wouldn't you know it?—a contestant stayed around for months and months. It seemed like forever. But this, it turned out, wasn't a bad thing at all. Ratings soared. *Jeopardy* had hatched its first celebrity.

His name was Ken Jennings. Nothing about the man suggested quiz show dominance. Unlike basketball, where a phenom like LeBron James emerged in high school, amid monster dunks, as the Next Big Thing, a *Jeopardy* champion like Jennings could surprise even himself. A computer programmer from Salt Lake City, Jennings had competed in quiz bowl events during high school and college. A turn on *Jeopardy* would be a kick. So in the summer of 2003, he and a friend drove from Salt Lake City to the *Jeopardy* studios in Culver City and took the qualifying exam. Jennings was pleased to pass it. And he was surprised, nine months later, to get the call that he'd been selected to play. He promptly started cramming his head with facts and dates about movies, kings, and presidents.

His first game came a month later. Before the game, Jennings, like every other contestant, had to tape a short promotion, a "Hometown Howdy," to be played in Salt Lake City the day before the show aired. It is typically corny, and his was no exception: "Hey there, Utah. This is Ken Jennings from Salt Lake City, and I hope the whole Beehive State will be buzzing about my appearance on *Jeopardy*." Little did he know that within months, not just the Beehive State, but the whole country, would be buzzing about Ken Jennings.

In his first game, he wrote in *Brainiac,* it was only through

the leniency of a judge's ruling that he managed to win. After two rounds, he held a slim $20,000 to $18,800 lead over
the next player, Julia Lazerus, a fundraiser from New York
City. The reigning champ, a Californian named Jerry Har-
vey, trailed far behind, with only $7,400. The category for Fi-
nal Jeopardy was The 2000 Olympics. Though Jennings had
been on his honeymoon during the two weeks of the Syd-
ney Olympics and hadn't seen a single event, he bet $17,201.
This would ensure victory if Lazerus bet everything and they
both got it right. If she wagered more modestly—betting that
he'd miss—and won, a wrong answer would cost Jennings the
game.

Trebek read the Final Jeopardy clue: "She's the first female track-and-field athlete to win five medals in five different events in a single Olympics." Jennings wrote that he was
racked by doubt. He knew that Marion Jones was the big
medal winner in that Olympics. (In 2007, Jones would ad-
mit to doping and surrender her medals.) To Jennings, Jones
seemed too obvious. Everyone knew her. There had to be
some kind of trick. But he couldn't come up with another
answer. In the end, following common *Jeopardy* protocol, he
skipped her first name and wrote: "Who is Jones?" A botched
first name or middle initial, players knew, turned a correct response into a wrong one. "Mary Jones" or "Marianne Jones"
would be incorrect. But a correct last name sufficed—or it
usually did. The trouble was that Jones was such a common
name, like Smith or Black, that someone who didn't know the
answer might have guessed it.

In the end, Jennings could have won the game by betting
nothing. Lazerus flubbed the clue, coming up only with "Who
is Gail?" a reference to Gail Devers, a gold medal sprinter in
the 1992 and 1996 games. She wagered $3,799, which left her

with $14,801. It was still more than enough to win if Jennings missed it or if the single name failed to satisfy the judges.

He showed his response: "Who is Jones?" Trebek paused and glanced at the judges. If there had been another prominent female track star named Jones, Jennings, like thousands of others, would have been a one-time loser on America's most popular quiz show. But the judges knew no other stars named Jones and approved his vague answer. "We'll accept that," Trebek said. Ken Jennings won the game and $37,201, becoming the new *Jeopardy* champion.

Millions of viewers witnessed the drama that June evening. Many of them probably figured that, like most champions, the skinny computer programmer who snuck through in Final Jeopardy would lose the next day or the day after that. In fact, by the time the "Jones" show aired, Jennings was already well into his streak. *Jeopardy* recorded its games two or three months ahead of time, and Friedman's team usually taped five games per day—a grueling ordeal for winning contestants. Between games, Trebek and the winner left the stage to change clothes, appearing ten minutes later with a new look—as if it were another day. Within an hour of his first victory, Jennings won again. In two days, he won his first eight games, then headed back to Salt Lake City. During his streak, he commuted between the two cities without disclosing what he was up to. Like all *Jeopardy* players, and even members of the studio audience, he had signed legal forms vowing not to disclose the results of the games before they aired. His streak was a secret.

As the weeks passed, the games seemed to become easier for him. He grew comfortable with the buzzer, could pick out the hints in the clues and read the signals of his mind. More often than not, Jennings did not just beat his competitors, he blew them away. After the first two rounds of a game, he had usually amassed more than twice the winnings of his nearest rival. This was known as a lock-out, for it rendered Final Jeopardy meaningless. As time passed, Jennings fell into a winning rhythm.

Millions of new viewers tuned into *Jeopardy* to see the summer sensation. In July, as Jennings extended his streak to thirty-eight games, ratings jumped 50 percent from those of the previous year, reaching a daily audience of fifteen million. *Jeopardy* rose to be the second-ranked TV show of the month, trailing only the CBS prime-time crime series *CSI*. In an added dividend for Friedman, *Jeopardy*'s rise also boosted ratings for its stablemate, *Wheel of Fortune,* which followed it on many channels.

Jennings, with his choirboy face and awkward grin, was a far cry from the tough guys on *CSI*. But he was proving to be a cognitive mauler. Some of his fallen opponents (who eventually numbered 148) took to calling themselves Road Kill and produced T-shirts for the growing club. Yet even while Jennings racked up wins he flashed humor, some of it even mischievous. One $200 clue in the category Tool Time read: "This term for a long-handled gardening tool can also mean an immoral pleasure-seeker." Jennings, his knowledge clearly extending into gangsta rap, responded: "What is a 'ho'?" That produced laughter and oohs and aahs from the crowd. A surprised Trebek struggled briefly for words, finally asking Jennings: "Is that what they teach you in school, in Utah?" His response was ruled incorrect. In fact, it could be argued that Jennings's gaffe was right—and far more clever than the intended answer ("What is a rake?"). He could have challenged the call, but he was so far ahead it was barely worth the bother.

What was so special about Ken Jennings? First, he knew

a lot. A practicing Mormon who had spent his childhood in Korea and had done missionary work in Spain, he knew the Bible and international affairs. He'd devoted himself to quiz bowls much of his life, the way others honed their skills in ice hockey or ballet, and he had a fabulous memory. Still, his peers considered him only an excellent player, not a once-in-a-generation phenomenon. "None of us who knew Ken saw this coming," said Greg Lindsay, a two-time *Jeopardy* champ who had crossed paths with Jennings in college quiz bowl tournaments.

Two things, according to his competitors, distinguished Jennings. First, he had an uncanny feel for the buzzer. This wasn't a mechanical ability but a uniquely human one. Sitting at the production table by the *Jeopardy* set, a game official waited for Trebek to finish reading the clue, then turned on the panel of lights on the big *Jeopardy* board. This signaled the opportunity to buzz. Players who buzzed too early got penalized: Their buzzers were locked out for a crucial quarter of a second, opening the door for others to buzz in. Jennings, said his competitors, had an almost magical feel for the rhythm of the buzzmeister. He anticipated the moment almost the way jazz musicians sense a downbeat. "Ken knew the buzzer," said Deirdre Basile, one of his early victims. "He had that down to a science."

His second attribute was a preternatural calm under pressure. Like other players, Jennings had a clear sense of what he knew. (This is known as "metacognition.") But knowing a fact is one thing, locating it quite another. People routinely botch the retrieval process, sometimes hunting for the name of a person standing right in front of them. This problem, known as "tip of the tongue syndrome," occurs more often when people are stressed—such as when they have less than four seconds to come up with an answer, thousands of dollars are at stake, and they're standing in front of a television audience of millions.

Bennett L. Schwartz, a psychologist at Florida International University, has studied the effects of emotion on tip of the tongue syndrome. He came up with questions designed to make people anxious, such as, "What was the name of the tool that executed people in the French Revolution?" With beheadings on their mind, he found, people were more likely to freeze up on the answer. Memory works on clues—words, images, or ideas that lead to the area where the potential answer resides. People suffering from tip of the tongue syndrome struggle to find those clues. For some people, Schwartz said, the concern that they might experience difficulty becomes a self-fulfilling prophecy. "I know the answer and I can't retrieve it," he said. "That's a conflict." And the brain appears to busy itself with this internal dispute instead of systematically trawling for the most promising clues and pathways. Researchers at Harvard, studying the brain scans of people suffering from tip of the tongue syndrome, have noted increased activity in the anterior cingulate—a part of the brain behind the frontal lobe, devoted to conflict resolution and detecting surprise.

Few of these conflicts appeared to interfere with Jennings's information retrieval. During his unprecedented seventy-four-game streak, he routinely won the buzz on more than half the clues. And his snap judgments that the answers were on call in his head somewhere led him to a remarkable 92 percent precision rate, according to statistics compiled by the quiz show's fans. This topped the average champion by 10 percent.

As IBM's scientists contemplated building a machine that could compete with the likes of Ken Jennings, they understood their constraints. Their computer, for all its power and speed, would be a first cousin of the laptops they carried around the Hawthorne lab. That was the technology at hand for a challenge in 2011. No neocortex, no neurons, no anterior cingulate, just a mountain of transistors etched into silicon processing ones and zeros. Any *Jeopardy* machine they built would struggle mightily to master language and common sense—areas that come as naturally to humans as breathing. Their machine would be an outsider. On occasion it would be clueless, even laughable. On the positive side, it wouldn't suffer from nerves. On certain clues it would surely piece together its statistical analysis and summon the most obscure answers with sufficient speed to match that of Ken Jennings. But could they ensure enough of these successes to win?

Ken Jennings's remarkable streak came to an end in a game televised in November 2004. Following a rare lackluster performance, he was only $4,400 ahead of Nancy Zerg, a real estate agent from Ventura, California. It came down to the Final Jeopardy clue: "Most of this firm's 70,000 seasonal white-collar employees work only four months a year."

The *Jeopardy* jingle came on, and Jennings put his brain into drive. But the answer, he said, just wasn't there. He didn't read the business pages of newspapers. Companies were one of his few weak spots. He guessed, "What is FedEx?" When Zerg responded correctly, "What is H&R Block?" Jennings knew his reign was over. During his streak, he had amassed more than $2.5 million in earnings and became perhaps the first national brand for general braininess since the disgraced Charles Van Doren.

Harry Friedman, of course, was far too smart a producer to let such an asset walk away. A year later, he featured Jennings in a wildly promoted Ultimate Tournament of Champions. This eventually brought Jennings into a threesome featuring the two leading money winners from before 2003, when winners were limited to five matches. Both Jerome Vered and Brad Rutter had retired as undefeated champions under the rules at the time. Rutter, who had dropped out of Johns Hopkins University and worked for a time at a music store in Lancaster, Pennsylvania, had never lost a *Jeopardy* match.

In the 2005 showdown, Rutter handled both Jennings and Vered with relative ease. He was so fast to the buzzer, Jennings later said, that sometimes the light to open the buzzing didn't appear to turn on. "It was off before it was on," he said. "I don't know if the filaments got warmed up." In the three days of competition, Rutter piled up 62,000, compared to 34,599 for Jennings and 20,600 for Vered. (These weren't dollars but points, since they were playing for a far larger purse.) Rutter won another $2 million, catapulting him past Jennings as the biggest money winner in *Jeopardy* history.

These two, Rutter and Jennings, were the natural competitors for an IBM machine. To establish itself as the *Jeopardy* king, the computer had to vanquish the best. These two players fit the bill. And they promised to be formidable opponents. They had human qualities a *Jeopardy* computer could never approach: fluency in language, an intuitive feel for hints and suggestion, and a mastery of ideas and concepts. Beyond that, they appeared to boast computer-like qualities: vast memories, fast processors, and nerves of steel. No tip-of-the-tongue glitches for Jennings or Rutter. But would a much-ballyhooed match against a machine awaken their human failings? Ferrucci and his team could always hope.

3

Blue J Is Born

IN THOSE EARLY DAYS of 2007, when Blue J was no more than a conditional promise given to Paul Horn, David Ferrucci harbored two conflicting fears. By nature he was given to worrying, and the first of his nightmare scenarios was perfectly natural: A *Jeopardy* computer would fail, embarrassing the company and his team.

But his second concern, failure's diabolical twin, was perhaps even more terrifying. What if IBM spent tens of millions of dollars and devoted centuries of researcher years to this project, played it up in the press, and then, perhaps on the eve of the nationally televised *Jeopardy* showdown, someone beat them to it? Ferrucci pictured a solitary hacker in a garage, cobbling together free software from the Web and maybe hitching it to Wikipedia and other online sites. What if the *Jeopardy* challenge turned out to be not too hard but too easy?

That would be worse, far worse, than failure. IBM would become the laughingstock of the tech world, an old-line company completely out of touch with the technology rev-olution—precisely what its corporate customers paid it billions of dollars to track. Ferrucci's first order of business was to make sure that this could never happen. "It was due diligence," he later said.

He had a new researcher on his team, James Fan, a young Chinese American with a fresh doctorate from the University of Texas. As a newcomer, Fan was free of institutional preconceptions about how Q-A systems should work. He had no history with the annual TRec competitions or IBM's Piquant system. Trim and soft-spoken, his new IBM badge hanging around his neck, Fan was an outsider. Unlike most of the team, based in New York or its suburbs, Fan lived with his parents in Parsippany, New Jersey, some seventy miles away. He was the closest thing Ferrucci had to a solitary hacker in a garage.

Fan, who emigrated as an eighteen-year-old from Shanghai to study at the University of Iowa and later Texas, had focused his graduate work on teaching machines to come to grips with our imprecise language. His system would help them understand, for example, that in certain contexts the symbol H_2O might represent a single molecule of water while in others it could refer to the sloshing contents of Lake Michigan. This expertise might eventually help teach a machine to understand *Jeopardy* clues and to hunt down answers. But it hardly prepared him for the job he now faced: building a *Jeopardy* computer all by himself. His system would be known as Basement Baseline.

As Fan undertook his assignment, Ferrucci ordered his small Q-A team to adapt their own system to the challenge, and he would pit the two systems against each other. Ferrucci called this "a bake-off." The inside team would use the Piquant technology developed at IBM while the outside team, consisting solely of James Fan, would scour the entire world for the data and software to jury-rig a bionic *Jeopardy* player. They each had four weeks and a set of five hundred *Jeopardy* clues to train on. Would either system be able to identify the parent bird of the roasted young squab (What is a pigeon?) or the sausage celebrated every year since 1953 in Sheboygan, Wisconsin (What is bratwurst?)? If so, would either have enough confidence in its answers to bet on them?

Ferrucci suspected at the time that his solitary hacker would come up with ideas that might prove useful. The bake-off, he said, would also send a message to the rest of the team that a *Jeopardy* challenge would require reaching outside the company for new ideas and approaches. He wanted to subject everyone to Darwinian pressures. The point was "to have technologies competing," he said. "If somebody's not getting it done, if he's stuck, we're going to take code away from him and give it to someone else." This, he added, was "horrific for researchers." Those lines of software may have taken months or even years to develop. They contained the researcher's ideas and insights reduced to a mathematical elegance. They were destined for greatness, perhaps coder immortality. And one day they could be ripped away and given to a colleague—a competitor, essentially—who might make better use of them. Not everyone appreciated this. "One guy went to his manager," Ferrucci said, "and said that the bake-off was 'bad for morale.' I said, 'Welcome to the WORLD!'"

So on a February day in 2007, James Fan set out to program a Q-A machine all by himself. He was relatively isolated in a second-floor office while the rest of Ferrucci's team mingled on the first floor. He would continue to run into them in the cafeteria, and they would attend meetings together. After all, they were colleagues, each one of them engaged in a venture that many in the company viewed as hopeless. "I was the most optimistic member of the team," Fan later said, "and I was thinking, 'We can make a decent showing.'" As he saw it, "decent" meant losing to human champions but nailing a few questions and ending up with a positive score.

Fan started by drawing up an inventory of the software tools and reference documents he thought he'd need for his machine. First would be a so-called type system. This would help the computer figure out if it was looking for a person, place, animal, or thing. After all, if it didn't know what it was looking for, finding an answer was little more than a crap-shoot; generating enough "confidence" to bet on that answer would be impossible. The computer would be lost.

For humans, distinguishing President George Washing-ton from the bridge named after him wasn't much of a chal-lenge. Context made it clear. Bridges didn't deliver inaugural addresses; presidents were rarely jammed at rush hour, with half-hour delays from New Jersey. What's more, when placed in sentences, people usually behaved differently than roads or bridges.

But what was simple for us involved hard work for a Q-A computer. It had to comb through the structure of the ques-tion, picking out the subjects, objects, and prepositions. Then it had to consult exhaustive reference lists that had been built up in the industry over decades, laying out hundreds of thou-sands of places, things, and actions and the web of relation-ships among them. These were known as "ontologies." Think of them as cheat sheets for computers. If a finger was a sub-ject, for example, it fell into human anatomy and was related to the hand and the thumb and to verbs such as "to point" and "to pluck." (Conversely, when "the finger" turned up as the object of the verb "to give," a sophisticated ontology might steer the computer toward the neighborhood of insults, gestures, and obscenities.)

In any case, Fan needed both a type system and a knowledge base to understand questions and hunt for answers. He didn't have either, so he took a hacker's shortcut and used Google and Wikipedia. (While the true *Jeopardy* computer would have to store its knowledge in its "head," prototypes like Fan's were free to search the Web.) From time to time, Fan found, if he typed a clue into Google, it led him to a Wikipedia page—and the subject of the page turned out to be the answer. The following clue, for example, would confound even the most linguistically adept computer. In the category The Author Twitters, it reads: "Czech out my short story 'A Hunger Artist'! Tweet done. Max Brod, pls burn my laptop." A good human *Jeopardy* player would see past the crazy syntax, quickly recognizing the short story as one written by Franz Kafka, along with a reference to Kafka's Czech nationality and his longtime associate Max Brod.

In the same way, a search engine would zero in on those helpful key words and pay scant attention to the sentence surrounding them. When Fan typed the clue into Google, the first Wikipedia page that popped up was "Franz Kafka," the correct answer. This was a primitive method. And Fan knew that a computer relying on it would botch the great majority of *Jeopardy* clues. It would be crashing and burning in the game against even ignorant humans, let alone Ken Jennings. But one or two times out of ten, it worked. For Fan, it was a start.

The month passed. Fan added more features to Basement Baseline. But at the end, the system was still missing vital components. Most important, it had no mechanism for gauging its level of confidence in its answers. "I didn't have time to build one," Fan said. This meant the computer didn't know what it knew. In a game, it wouldn't have any idea when to buzz. Fan could conceivably have programmed it with simple rules. It could be instructed to buzz all the time—a serious money loser, considering it flubbed two clues for every one it got right. Or he could have programmed it to buzz in every category in which it got the first clue right. That would signal that it was oriented to the category. But his machine didn't have any way to learn that its response was right or wrong. It lacked a feedback loop. In the end, Fan blew off game strategy entirely and focused simply on building a machine that could answer *Jeopardy* clues.

It soon became clear that the bake-off, beyond a test of technologies, also amounted to a theater production staged by David Ferrucci. It was tied to inside politics. Ferrucci didn't believe that the Piquant platform could ever be adapted to *Jeopardy*. It wasn't big or robust enough. Yet there were expectations within the company that Piquant, which represented more than twenty researcher years, would play an important role. To build the far bigger machine he envisioned, Ferrucci needed to free himself, and the project, from the old guard's legacy. For this, Piquant had to fail. He didn't spell this out. But he certainly didn't give the team the guidance, or the time, to overhaul the system. So besides training the machine on five hundred *Jeopardy* clues and teaching it to answer them in the form of questions, the Piquant team left the system largely unchanged. "You could have guessed from the outset that the success rate was not going to be very high," said Jennifer Chu-Carroll, a member of the team. Piquant was being led to a public execution.

The bake-off took place on a March morning at the Hawthorne lab. The results, from Ferrucci's perspective, were ideal. The Piquant system succeeded on only 30 percent of the clues, far below the level needed for *Jeopardy*. It had high confidence on only 5 percent of them, and of those it got only 47 percent right. Fan's Basement Baseline fared almost as well by a number of measures but was still woefully short of what was needed. Fan proved that a hacker's concoction was far from *Jeopardy* standards—which was a relief. But by nearly matching the company's state-of-the-art in Q-A technology, he highlighted its inadequacies.

The *Jeopardy* challenge, it was clear, would require another program, another technology platform, and a far bolder approach. Ferrucci wouldn't hesitate to lift algorithms and ideas from both Piquant and Basement Baseline, but the project demanded far more than a recasting of IBM technologies. It was too big for a single company, even one as burly as IBM. The Blue J machine, Ferrucci said, would need "the most sophisticated intelligence architecture the world has ever seen." For this, the *Jeopardy* team would have to reach out to the universities doing the most exciting work in AI, including MIT, Carnegie Mellon, and the University of Texas. "We needed all the brains we could get behind this project," he said.

Back in the late '70s, when he was commuting from the Bronx to his high school in suburban New Rochelle, Ferrucci and his best friend at the time, Tony Marciano, had an idea for a new type of machine. They called it "a reverse-dictionary." The idea, Ferrucci said, was to build a machine that could find elusive words. "You know how it is when you want to express something, but you can't think of the right word for it? A dictionary doesn't help at all, because you don't know what to look up. We wanted to build the machine that would give you the word." This was before they'd ever seen a computer. "We were thinking of a mechanical thing."

It sounded like a thesaurus. But Ferrucci bridled a bit at the suggestion that his dream machine had existed for centuries as a book. "No, you don't give it the synonyms, just the definition," he said. "Basically we were scratching this idea that the computer could understand your meaning, your words, your definitions, and could come up with the word."

Ferrucci was a hot shot at science at Iona Grammar School, a Catholic boys school. He and Marciano—who, according to Ferrucci, "did calculus on his cuff links"—regarded even their own brains as machines. Marciano, for example, had the idea that devoting brain space to memory storage was wasteful. It distracted neurons from the more important work of processing ideas. So when people asked him questions requiring recall, he would respond, "Ask Dave. He's willing to use memory."

Ferrucci's father, Antonio, had come to the United States from Italy after the Second World War. He had studied some law and engineering in the dying days of Mussolini's regime, but he arrived in New York without a profession and ended up driving trucks and working in construction. He and Ferrucci's mother, Connie, wanted their son to be a doctor. One summer during high school, Ferrucci had planned just to hang around with his friends and "play." His father wouldn't stand for it. "He'd gotten something in the mail about a math and computer course at Iona College. He says, 'You've got the grades, why don't you sign up for that?'"

At Iona, Ferrucci came face-to-face with his first computer. It featured a hulking cathode ray tube with a black screen and processed data encoded on teletype. He fell for it immediately. "Here was a machine," he said. "You told it to do stuff, and it *did* what you told it. I thought, 'This is *big*.' I called up Tony Marciano, and I said, 'You get your butt over here, into this room at Iona College. You've got to see this machine.'"

Marciano, who later studied computer science and went on to become a finance professor at New York University's Stern Business School, met Ferrucci later that afternoon. The two of them stayed long into the evening, paging through a single manual, trying out programs on the computer and getting the machine to spit out differential equations. At that point, Ferrucci knew that he wanted to work with computers. However, he didn't consider it a stand-alone career. A computer was a tool, as he saw it, not a destination. Anyway, he was going to be a doctor.

He went on to Manhattan College, a small Catholic school that was actually in The Bronx, a few miles north of Manhattan. There he followed the pre-med track as a biology major and took computer science on the side. "I did a bunch of programming for the physiology lab," he said. "Everything I did in biology I kept relating to computers." The way technology was advancing, it seemed, there had to be a place for computers in medicine.

One night, Ferrucci was taking a practice exam in a course for the MCATs, the Medical College Admission Tests. "I was with all my pre-med friends," he said. "This is midway through the course. The proctor says, 'Open to page 10 and start taking the sample chemistry test.' I opened it up and I started doing the questions, and all of a sudden I said, 'You know what? I'm not going to be a doctor!' And I closed the test and I went up to the proctor and I said, 'I'm quitting. I don't want to be a doctor.' He said, 'You're not going to get your $500 back.' I said, 'Whatever.'"

Ferrucci left the building and made two phone calls. He dialed the easier one first, telling his girlfriend that he'd just walked out of the MCAT class and was giving up on medicine. Then he called his father. "That was a hard call to make," Ferrucci said. "He was very upset in the beginning."

His MCAT insight, while steering him away from medicine, didn't put him on another clear path. He still didn't know what to do. "I started looking for graduate programs in physiology that had a strong computing component," he said. "After about a week or two of that, I suddenly said, 'Wait a minute.'" He called this his "second-level epiphany." He asked himself why he was avoiding the obvious. "I was really interested in the computer stuff," he said, "not the physiology. So I'd have to make a complete break." He applied to graduate school in computer science and went upstate, to Rensselaer Polytechnic Institute (RPI), in Troy, New York.

In his first stint at IBM Research, between getting his master's and his doctorate at RPI, Ferrucci delved into AI. By that time, in the late '80s, the industry had split into two factions. While some scientists still pursued the initial goal of thinking machines, or general intelligence, others looked for more focused applications that could handle real jobs (and justify the research). The king of "narrow AI," and Ferrucci's focus, was the expert system. The idea was to develop smart software for a specific industry. A program designed, say, for the travel industry could answer questions about Disneyland or Paris, find cheap flights, and book hotels. These specialists wouldn't have to puzzle out the context of people's conversations. The focus of their domains would make it clear. For that electronic expert in travel, for example, "room" would mean only one thing. The computer wouldn't have to concern itself with "room" in the backseat of a Cadillac or "room" to explore in the undergraduate curriculum at Bryn Mawr. If it were asked about such things, it would draw a blank. Computers that lacked range and flexibility were known as brittle. The one-trick ponies seen as expert systems almost defined the term. Many in the industry didn't consider them AI at all. They certainly didn't think or act like people.

To build a more ambitious-thinking machine, some looked to the architecture of the human brain. Indeed, while Ferrucci was grappling with expert systems, other researchers were piecing together an altogether different species of program, called "neural networks." The idea had been bouncing around at least since 1948, when Alan Turing outlined it in a paper called "Intelligent Machinery." Like much of his thinking, Turing's paper was largely theoretical. Computers in his day, with vacuum tubes switching the current on and off, were too primitive to handle such work. (He died in 1954, the year that Texas Instruments produced the first silicon transistor.) However, by the '80s, computers were up to the job. Based on rudimentary models of neurons, these networks analyzed the behavior of complex systems, such as financial markets and global weather, and used statistical analysis to predict how they would behave over time.

A neural network functioned a bit like a chorus. Picture a sing-along concert of Handel's *Messiah* in Carnegie Hall. Some five thousand people show up, each one wearing a microphone. You play the music over loudspeakers and distribute musical scores. That's the data input. Most of the people start singing while others merely hum or chat with their neighbors. In a neural net, the learning algorithm picks out the neurons that appear to be replicating the pattern, and it gives them more sway. This would be like turning up the microphones of the people who are singing well, turning down the mikes of those who sing a tad off key—and shutting out the chatterers altogether. The net focuses not only on the individuals but on the connections among them. In this analogy, perhaps the singers start to pay attention to one another and organize, the tenors in one section, sopranos in another. By the end of a long training process, the Carnegie Hall network both interprets the data and develops an expertise in Handel's motifs and musical structure. The next week, when the music switches to Gershwin, new patterns emerge. Some of the chatterers, whose mikes were turned off, become stars. With time, this assemblage can identify new pieces of music, recognizing similar themes and variations. And the group might even set off an alarm if the director gets confused and starts playing Vivaldi instead of Handel.

Neural networks learned, and even evolved. In that sense, they crudely mimicked the human brain. People driving cars, for example, grow to respond to different patterns—the movement of traffic, the interplay between the wheel and the accelerator—often without thinking. These flows are reflected by neural connections in the brain, lots of them working in parallel. They're reinforced every time an experience proves their usefulness. But a change, perhaps a glimpse of a cyclist riding against traffic, snaps them from their reverie. In much the same way, neural networks became very good at spotting anomalies. Credit card companies began to use them to note unexpected behavior—an apparent teetotaler buying $500 of Finnish vodka or a frugal Nebraskan renting luxury suites in Singapore. Various industries, meanwhile, used neural networks to look ahead. As long as the future stayed true to the past—not always a safe assumption, as any mortgage banker can attest—they could make solid predictions.

Unlike the brittle expert systems, neural networks were supple. They specialized in pattern detection, not a series of if/then commands. They never choked on changes in the data but simply adjusted. While expert systems processed data sequentially, as if following a recipe, the electronic neurons crunched in unison—in parallel. Their weakness? Since these collections of artificial neurons learned by themselves, it was nearly impossible to figure out how they reached their conclusions or to understand what they were picking up about the world. A neural net was a black box.

By the time Ferrucci returned to IBM Research, in 1995, he was looking beyond expert systems and neural nets. In his spare time, he and a colleague from RPI, Selmer Bringsjord, were building a machine called Brutus, which wrote fiction. And they were writing a book about their machine, *Artificial Intelligence and Literary Creativity.* Brutus, they wrote, is "utterly devoid of emotion, but he nonetheless seems to have within his reach things that touch not only our minds, but our heart."

The idea for the program, Ferrucci later said, came when Bringsjord asked him if a machine could create its own story line. Ferrucci took up the challenge. Instead of teaching the machine to dream up plots, he programmed it with about a dozen themes, from betrayal to revenge. For each theme, the machine was first given a series of literary examples and then a program to develop stories along those lines. One of its models for betrayal was Shakespeare's *Julius Caesar* (the program was named for Caesar's confidant-turned-conspirer, Brutus). The program produced serviceable plots, but they were less than riveting. "The one thing it couldn't do is figure out if something was interesting," Ferrucci said. "Machines don't understand that."

In his day job, Ferrucci was teaching computers more practical lessons. As head of Semantic Analysis and Integration at IBM, he was trying to instruct them to make sense of human communication. On the Internet, records of our words and activities were proliferating as never before. Companies—IBM and its customers alike—needed tools to interpret these new streams of information and put them to work. Ideally, an IBM program would tell a manager what customers or employees were saying or thinking as well as what trends and insights to draw from them and perhaps what decisions to make.

Within IBM itself, some two hundred researchers were developing a host of technologies to mine what humans were writing and saying. But each one operated within its own specialty. Some parsed sentences, analyzing the grammar and vocabulary. Others hunted Google-style for keywords and Web links. Some constructed massive databases and ontologies to organize this knowledge. A number of them continued to hone expert systems and neural networks. Meanwhile, the members of the Q-A team coached their computer for the annual TRec competitions. "We had lots of different pockets of researchers working on these different analytical algorithms," Ferrucci said. "But any time you wanted to combine them, you had a problem." There was simply no good way to do it.

In the early 2000s, Ferrucci and his team put together a system to unify these diverse technologies. It was called UIMA, Unstructured Information Management Architecture. It was tempting to think of UIMA as a single brain and all of the different specialties, from semantic analysis to fact-checking, as cognitive regions. But Ferrucci maintained that UIMA had no intelligence of its own. "It was just plumbing," he said. Idle plumbing, in fact, because for years it went largely unused.

But a *Jeopardy* project, he realized, could provide a starring role for UIMA. Blue J would be more than a single machine. His team would pull together an entire conglomeration of Q-A approaches. The machine would house dozens, even hundreds of algorithms, each with its own specialty, all of them chasing down answers at the same time. A couple of the jury-rigged algorithms that James Fan had ginned up could do their thing. They would compete with others. Those that delivered good answers for different types of questions would rise in the results—a bit like the best singers in the Handel sing-along. As each one amassed its record, it would gain stature in its specialty and be deemed clueless in others. Loser algorithms—those that failed to produce good results in even a single niche—would be ignored and eventually removed. (Each one would have to prove its worth in at least one area to justify its inclusion.) As the system learned which algorithms to pay attention to, it would grow smarter. Blue J would evolve into an ecosystem in which the key to survival, for each of the algorithms, would be to contribute to correct responses to *Jeopardy* clues.

While part of his team grappled with Blue J's architecture, Ferrucci had several researchers trolling the Internet for *Jeopardy* data. If this system was going to compete with humans in the game, it would require two types of information. First, it needed *Jeopardy* clues, thousands of them. This would be the machine's study guide—what those in the field of machine learning called a training set. A human player might watch a few *Jeopardy* shows to get a feel for the types of clues and then take some time to study country capitals or brush up on Shakespeare. The computer would do the same work statistically. Each *Jeopardy* clue, of course, was unique and would never be repeated, so it wasn't a question of learning the answers. But a training set would orient the researchers. Given thousands of clues, IBM programmers could see what percentage of them dealt with geography, U.S. presidents, words in a foreign language, soap operas, and hundreds of other categories—and how much detail the computer would need for each. The clue asking which presidential candidate carried New York State in 1948, for example ("Who is Thomas Dewey?"), indicated that the computer would have to keep track of White House losers as well as winners. What were the odds of a presidential loser popping up in a clue?

Digging through the training set, researchers could also rank various categories of puzzles and word games. They could calculate the odds that a *Jeopardy* match would include a puzzling Before & After, asking, for example, about the "Kill Bill star who played 11 seasons behind the plate for the New York Yankees" ("Who is Uma Thurman Munson?"). A rich training set would give them a chance to scrutinize the language in *Jeopardy* clues, including abbreviations, slang, and foreign words. If the machine didn't recognize AKA as "also known as" or "oops!" as a misunderstanding, if it didn't recognize "sayonara," "au revoir," "auf Wiedersehen," and hundreds of other expressions, it could kiss entire *Jeopardy* categories goodbye. Without a good training set, researchers might be filling the brain of their bionic student with the wrong information.

Second, and nearly as important, they needed data on the performance of past *Jeopardy* champs. How often did they get the questions right? How long did they take to buzz in? What were their betting strategies in Double Jeopardy and Final Jeopardy? These humans were the competition, and their performance became the benchmark for Blue J.

In the end, it didn't take a team of sleuths to track down much of this data. With a simple Internet search, they found a Web site called J! Archive, a trove of historical *Jeopardy* data. A labor of love by *Jeopardy* fans, the site detailed every game in the show's history, with the clues, the contestants, their answers—and even the comments by Alex Trebek. Here were more than 180,000 clues, hundreds of categories, and the performance of thousands of players, from first-time losers to champions like Brad Rutter and Ken Jennings.

In these early days, the researchers focused only on Jennings. He was the gold standard. And with records of his seventy-four games—more than four times as many as any other champion—they could study his patterns, his strengths and vulnerabilities. They designed a chart, the Jennings Arc, to map his performance: the percentage of questions on which he won the buzz and his precision on those questions. Each of his games was represented by a dot, and the best ones, with high buzz and high accuracy, floated high on the chart to the extreme right. His precision averaged 92 percent and occasionally reached 100 percent. He routinely dominated the buzz, on one game answering an astounding 75 percent of the clues. For each of these games, the IBM team calculated how well a competitor would have to perform to beat him. The numbers varied, but it was clear that their machine would need to win the buzz at least half the time, get about nine of ten right—and also win its share of Daily Doubles.

In the early summer of 2007, after the bake-off, the *Jeop-*

ardy team marked the performance of the Piquant system on the Jennings Arc. (Basement Baseline, which lacked a confidence gauge, did not produce enough data to be charted there.) Piquant's performance was so far down and to the left of Ken Jennings's dots, it appeared to be . . . well, exactly what it was: an alien species—and not destined for *Jeopardy* greatness.

When word of this performance spread around the Yorktown labs, it only fueled the concerns that Ferrucci's team was heading for an embarrassing fall—if it ever got that far. Mark Wegman, then the head of computer science at IBM Research, described himself as someone who's "usually wildly optimistic about technology." But when he saw the initial numbers, he said, "I thought there was a 10 percent chance that in five years we could pull it off."

For Ferrucci, Piquant's failure was anything but discouraging. It gave him the impetus to march ahead on a different path, toward Blue J. "This was a chance to do something really, really big," he said. However, he wasn't sure his team would see it this way. So he gathered the group of twelve in a small meeting room at the Hawthorne labs. He started by describing the challenges ahead. It would be a three- to five-year project, similar in length to a military deployment. It would be intense, and it could be disastrous. But at the same time they had an opportunity to do something memorable. "We could sit here writing papers for the next five years," he said, "or we build an entirely new type of computer." He introduced, briefly, a nugget of realpolitik. There would be no other opportunities for them in Q-A technologies within IBM. He had effectively engineered a land grab, putting every related resource into his *Jeopardy* ecosystem. If they wanted to do this kind of science, he said, "this was the only place to be."

Then he went around the room with a simple question: "Are you in or are you out?"

One by one, the researchers said yes. But their response was not encouraging. The consensus was that they could build a machine that could compete—but probably not beat—a human champion. "We thought it could earn positive money before getting to Final Jeopardy," said Chu-Carroll, one of the only holdovers on Ferrucci's team from the old TRec unit. "At least we wouldn't be kicked off the stage."

With this less than ringing endorsement, Ferrucci sent word to Paul Horn that the *Jeopardy* challenge was on. He promised to have a machine, within twenty-four months, that could compete against average human players. Within thirty-six to forty-eight months, his machine, he said, would beat champions one-quarter of the time. And within five to seven years, the *Jeopardy* machine would be "virtually unbeatable." He added that this final goal might not be worth pursuing. "It is more useful," he said, "to create a system that is less than perfect but easily adapted to new areas." A week later, Ferrucci and a small team from IBM Research flew to Culver City, to the Robert Young Building on the Sony lot. There they'd see whether Harry Friedman would agree to let the yet-to-be-built Blue J play *Jeopardy* on national television.

4

Educating Blue J

JENNIFER CHU-CARROLL, sitting amid a clutter of hardware and piles of paper in her first-floor office in the Hawthorne labs, wondered what in the world to teach Blue J. How much of the Bible would it have to know? The Holy Book popped up in hundreds of *Jeopardy* clues. But did that mean the computer needed to know every psalm, the laws of Deuteronomy, Jonah's thoughts and prayers while inside the whale? Would a dose of Dostoevsky help? She could feed it *The Idiot, Crime and Punishment,* or any of the other classics that might pop up in a *Jeopardy* clue. When it came to traditional book knowledge, feeding Blue J's brain was nearly as easy as Web surfing.

This was in July 2007. Chu-Carroll's boss, David Ferrucci, and the small IBM contingent had just flown back from Culver City, where they had been given a provisional thumbs-up from Harry Friedman. A man-machine match would take place, perhaps in late 2010 or early 2011. IBM needed the deadline to mobilize the effort within the company and to establish it as a commitment, not just a vague possibility. *Jeopardy*, for its part, would bend the format a bit for the machine. The games would not include audio or visual clues, where contestants have to watch a snippet of video or recognize a bar of music. And they might let the machine buzz electronically instead of hitting a physical button. The onus, according to the preliminary agreement, was on IBM to come up with a viable player in time for the match.

It was up to Chu-Carroll and a few of her colleagues to map out the machine's reading curriculum. Chu-Carroll had black bangs down to her eyes and often wore sweatshirts and jeans. Like practically everyone else on the team, she had a doctorate in computer science, hers from the University of Delaware. She had worked for five years at Lucent Technology's Bell Labs, in New Jersey. There she taught machines how to participate in a dialogue and how to modulate their voices to communicate different signals. (Lucent was developing automated call centers.) When Chu-Carroll came to IBM in 2001, joining her husband, Mark, she plunged into building Q-A technologies. (Mark later left for Google.)

In mid-2007, the nascent *Jeopardy* system wasn't really a machine at all. Fragments of a *Jeopardy* player existed as a collection of software programs, some of them hand-me-downs from the recent bake-off, all of them easy to load onto a laptop. As engineers pieced together an architecture for the new system, Chu-Carroll pondered a fundamental question: How knowledgeable did this computer really need to be? One of its forebears, Basement Baseline, had hunted down its answers on the Web. Blue J wouldn't have that luxury. So as Chu-Carroll sat down for Blue J's first day of school, her pupil was a tabula rasa.

She quickly turned to a promising resource. James Fan had already demonstrated the value of Wikipedia for answering a small subsection of *Jeopardy* clues. "It related to popular culture and what people care about," Chu-Carroll said. So she set to work extracting much of the vast corpus of Wikipedia articles from the online site and putting them into a format that Blue J could read.

But how about books? Gutenberg.org offered hundreds of classics for free, along with a ranking of the most popular downloads. Chu-Carroll could feed any or all of them to Blue J. After all, words didn't take up much space. *Moby Dick,* for example, was only 1.5 megabytes. Photographs on new camera phones packed more bits than that. So one day she downloaded the Gutenberg library and gave Blue J a crash course on the Great Books.

"It wasn't a smart move," she later admitted. "One of the most popular books on Gutenberg was a manual for surgeons from a hundred years ago." This meant that when faced with a clue about modern medicine, Blue J could be consulting a source unschooled in antibiotics, CAT scans, and HIV, one fixated instead on scurvy, rickets, and infections (not to mention amputations) associated with trench warfare. "I'm not sure why that book is so popular," said Chu-Carroll. "Are people doing at-home surgery?"

Whatever their motives, most human readers knew exactly what they were getting when downloading the medical relic. If not, they quickly found out. In addition to surgical descriptions, the book contained extraordinary pictures of exotic and horrifying conditions, such as elephantiasis of the penis. Aside from these images, the book's interest was largely historical. Humans had little trouble placing it in this context. Chu-Carroll's pupil, by contrast, had a maddening habit endemic among its ilk: It tended to take every source at its word.

Blue J's literal-mindedness posed the greatest challenge at every step of its education. Finding suitable data for this gullible machine was only the first task. Once Blue J had its source material, from James Joyce to archives of the Boing-Boing blog, the IBM team would have to teach it to make sense of all those words: to place names and facts into context and to come to grips with how they were related to one another. Hamlet, to pick one example, was related not only to his mother, Gertrude, but also to Shakespeare, Denmark, Elizabethan literature, a famous soliloquy, and themes ranging from mortality to self-doubt. Preparing Blue J to navigate all of these connections for virtually every entity on earth, factual or fictional, would be the machine's true education. The process would involve creating, testing, and fine-tuning thousands of algorithms. The final challenge would be to prepare it to play the game itself. Blue J would have to come up with answers it could bet on within three to five seconds. That job was still a year or two down the road.

For now, Chu-Carroll found herself contemplating academic heresy. Like college students paging through Cliff's Notes or surfing Wikipedia, she began to wonder whether Blue J should bother with books at all. Each one contained so many passages that could be misconstrued. In the lingo of her colleagues, books had a sky-high noise-to-signal ratio. The signals, the potential answers, swam in oceans of words, so-called noise.

Imagine Blue J reading Mark Twain's *Huckleberry Finn*. In one section, Huck and the escaped slave, Jim, are contemplating the night sky:

We had the sky up there, all speckled with stars, and we used to lay on our backs and look up at them, and discuss about whether they was made or only just happened. Jim he allowed they was made, but I allowed they happened; I judged it would have took too long to MAKE so many. Jim said the moon could a LAID them; well, that looked kind of reasonable, so I didn't say nothing against it, because I've seen a frog lay most as many, so of course it could be done.

Assuming that Blue J could slog through the idiomatic language—no easy job for a computer—it could "learn" something about the cosmos. Both characters, it appeared, agreed that the moon, like a frog, could have laid the stars. It seemed "reasonable" to them, a conclusion Blue J would be likely to respect. A human would put that passage into context, learn something about Jim and Huck, and perhaps laugh. Blue J, it was safe to say, would never laugh. It would likely take note of an utterly fallacious parent-offspring relationship between the moon and the stars and record it. No doubt its mad hunt through hundreds of sources to answer a single *Jeopardy* clue would bring in much more astronomical data and statistically overwhelm this passage. In time, maybe the machine would develop trusted sources for such astronomical questions and wouldn't be so foolish as to consult Huck Finn and Jim about the cosmos. But still, most books had too many words—too much noise—for the job ahead.

This led to an early conclusion about a *Jeopardy* machine. It didn't need to know books, plays, symphonies, or TV sitcoms in great depth. It only needed to know *about* them. Unlike literature students, the machine would not be pressed to compare and contrast the themes of family or fate in *Hamlet* with those in *Oedipus Rex*. It just had to know they were there. When it came to art, it wouldn't be evaluating the brushwork of Velázquez and Manet. It only needed to know some basic biographical facts about them, along with a handful of their most famous paintings. Ken Jennings, Ferrucci's team learned, didn't prepare for *Jeopardy* by plowing through big books. In *Brainiac,* he described endless practice with flash cards. The conclusion was clear: The IBM team didn't need a genius. They had to build the world's most impressive dilettante.

From their statistical analysis of twenty thousand *Jeopardy* clues drawn randomly from the past twenty years, Chu-Carroll and her colleagues knew how often each category, from U.S. presidents to geography, was likely to pop up. Cities and countries each accounted for a bit more than 2 percent of the clues; Shakespeare and Abraham Lincoln were regulars on the big board. The team proceeded to load Blue J with the data most likely to contain the answers. It was a diet full of lists, encyclopedia entries, dictionaries, thesauruses, newswire articles, and downloaded Web pages. Then they tried out batches of *Jeopardy* clues to see how it fared.

Blue J was painfully slow. Laboring on a single computer, it created a logjam of data, sending information through the equivalent of a skinny straw when it needed a fire hose. It didn't have enough bandwidth (the rate of data transfer). And it lacked computing muscle. This led to delays, or latency. It often took an hour or two to puzzle out the meaning of the clue, dig through its data to come up with a long list of possible answers, or "candidates," evaluate them, choose one, and decide whether it was confident enough to bet. The best way to run the tests, Chu-Carroll and her colleagues eventually realized, was to ask Blue J questions before lunch. The machine would cogitate; they'd eat. It was an efficient division of labor.

An hour or so after they returned to the office, Blue J's list of candidate answers would arrive. Many of them were on target, but some were ridiculous. One clue, for example, read: "A 2000 ad showing this pop sensation at ages 3 & 18 was the 100th "got milk?" ad." Blue J, after exhaustive work, missed the answer ("Who is Britney Spears?") by a mile, suggesting "What is Holy Crap?" as a possibility. The machine also volunteered that the diet of grasshoppers was "kosher" and that the Russian word for goodbye ("What is *do svidaniya*?") was "cholesterol."

The dumb answers didn't matter—at least not yet. They had to do with the computer's discernment, which was still primitive. The main concern for the *Jeopardy* team at this stage was whether the correct answer popped up anywhere at all on its list. This was its measure of binary recall, a metric that Blue J shared with humans. If a student in geography were asked about the capital of Romania, she might come up with Budapest and Bucharest and not remember which of the two was right. In Blue J's world, those would be her candidate answers, and she clearly had the data in her head to answer correctly. From Chu-Carroll's perspective, this student would have passed the binary recall test and needed no more data on capitals. (She just had to brush up on the ones she knew.) In a similar test, Blue J clearly didn't recognize Britney Spears or *do svidaniya* as the correct answer. But if those words showed up somewhere on its candidate lists, then it had the wherewithal to answer them (once it got smarter). By focusing on categories where Blue J struck out most often, the researchers worked to fill in the holes in its knowledge base.

Week by week, gigabyte by gigabyte, Blue J's trove of data grew. But by the standards of consumer electronics, it remained a pip-squeak. The fifth-generation iPods, which were selling five miles down the road, stored 30 to 80 gigabytes.

Blue J topped out at the level of a midrange iPod, at some 75 gigabytes. Within a year or two, cell phones might hold as much. But there was a reason for the coziness of Blue J's stash. The more data it had to rummage through, the longer it took—and it was already painfully slow. What's more, the easiest and clearest sources for Blue J to digest were lists and encyclopedia entries. They were short and to the point, and harder for their literal-minded pupil to misinterpret. And they didn't take up much disk space.

While Chu-Carroll wrestled with texts, James Fan was part of a team grappling with a challenge every bit as important: coaxing the machine to understand the convoluted *Jeopardy* clues. If Blue J couldn't figure out what it was supposed to look for, the data on its disk wouldn't make any difference. From Blue J's perspective, each clue was a riddle to be decoded, and the key was to figure out the precise object of its hunt. Was it a country? A person? A kind of fish? In *Jeopardy*, it wasn't always clear.

The crucial task was to spot the word representing what Blue J was supposed to find. In everyday questions, finding it was simple. In "Who assassinated President Lincoln?" or "Where did Columbus land in 1492?" the "who" and "where" point to a killer and a place. But in *Jeopardy*, where the clues are statements and the answers questions, finding these key words, known as Lexical Answer Types (LATs), was a lot trickier. Often a clue would signal the LAT with a "this," as in: "*This* title character was the crusty & tough city editor of the *Los Angeles Tribune*." Blue J had no trouble with that one. It identified "title character" as its focus and returned from its hunt with the right answer ("Who is Lou Grant?").

But others proved devilishly hard. This clue initially left Blue J befuddled: "In nine-ball, whenever you sink this, it's a scratch." Blue J, Fan said, immediately identified "this" as the object to look for. But what was "this?" The computer had to analyze the rest of the sentence. "This" was something that sank. But it was not related, at least in any clear way, to vessels, the most common sinking objects. To identify the LAT, Blue J would have to investigate the two other distinguishing words in the clue, "nine-ball" and "scratch." They led the computer to the game of pool and, eventually, to the answer ("What is a cue ball?").

Sometimes the LAT remained a complete mystery. The computer, Fan said, had all kinds of trouble figuring out what to look for in this World Leaders clue: "In 1984, his grandson succeeded his daughter to become his country's prime minister." Should the computer look for the grandson? The daughter? The country? Any human player would quickly understand that it was none of the above. The trick was looking for a single person whose two roles went unmentioned: a father and a grandfather. To unearth this, Blue J would have had to analyze family relationships. In the end, it failed, choosing the grandson ("Who is Rajiv Gandhi?"). In its list of answers, Blue J did include the correct name ("Who is Nehru?"), but it had less confidence in it.

Troubles with specific clues didn't matter. Even Ken Jennings only won the buzz 62 percent of the time. Blue J could afford to pass on some. The important thing was to fix chronic mistakes and to orient the machine to succeed on as many clues as possible. In previous weeks, Fan and his colleagues had identified twenty-five hundred different LATs in *Jeopardy* clues and ranked them by their frequency. The easiest for Blue J were the most specific. The machine could zero in on songs, kings, criminals, or plants in a flash, but most of them were more vague. "He," for example, was the most common, accounting for 2.2 percent of the clues. Over the coming months, Fan would have to teach Blue J how to explore the rest of each clue to figure out exactly what kind of "he" or "this" it should look for.

It was possible, Ferrucci thought, that someday a machine would replicate the complexity and nuance of the human mind. In fact, in IBM's Almaden research labs, on a California hilltop high above Silicon Valley, a scientist named Dharmendra Modha was building a simulated brain boasting seven hundred million electronic neurons. Within years, he hoped to map the brain of a cat, then a monkey, and eventually a human. But mapping the human brain, with its hundred billion neurons and trillions or quadrillions of connections among them, was a long-term project. With time, it might result in a bold new architecture for computing that would lead to a new level of computer intelligence. Perhaps then machines would come up with their own ideas, wrestle with concepts, appreciate irony, and think more like humans.

But such a machine, if it was ever built, would not be ready for Ferrucci. As he saw it, his team had to produce a functional *Jeopardy* machine within two years. If Harry Friedman didn't see a viable machine by 2009, he would never greenlight the man-machine match for late 2010 or early 2011. This deadline compelled Ferrucci and his team to assemble their machine with existing technology—the familiar silicon-based semiconductors, servers whirring through billions of calculations and following instructions from lots of software programs that already existed. In its guts, Blue J would not be so different from the ThinkPad Ferrucci lugged from one meeting to the next. Its magic would have to come from its massive scale, inspired design, and carefully tuned algorithms. In other words, if Blue J became a great *Jeopardy* player, it would be less a breakthrough in cognitive science than a triumph of engineering.

Every computing technology Ferrucci had ever touched, from the first computer he saw at Iona to the Brutus machine that spit out story plots, had a clueless side to it. Such machines could follow orders and carry out surprisingly complex jobs. But they were nowhere close to humans. The same was true of expert systems and neural networks: smart in one area, dumb in every other. And it was also the case with the *Jeopardy* algorithms his team was piecing together in the Hawthorne labs. These sets of finely honed computer commands each had a specialty, whether it was hunting down synonyms, parsing the syntax of a clue, or counting the most common words in a document. Beyond these meticulously programmed tasks, each was helpless.

So how would Blue J concoct broader intelligence—or at least enough of it to win at *Jeopardy*? Ferrucci considered the human brain. "If I ask you '36 plus 43,' a part of you goes, 'Oh, I'll send that question over to the part of my brain that deals with math,'" he said. "And if I ask you a question about literature, you don't stay in the math part of your brain. You work on that stuff somewhere else." Now this may be the roughest approximation of how the brain works, but for Ferrucci's purposes, it didn't matter. He knew that the brain had different specialties, that people instinctively skipped from one to another, and that Blue J would have to do the same thing.

Unlike a human, however, Blue J wouldn't know where to start. So with its vast resources, it would start everywhere.

Instead of reading a clue and assigning the sleuthing work to specialist algorithms, Blue J would unleash scores of them on a hunt, then see which one came up with the best answer. The algorithms inside Blue J—each following a different set of marching orders—would bring in competing results. This process, a lot less efficient than the human brain, would require an enormous complex of two thousand processors, each handling a different piece of the job.

To see how these algorithms carried out their hunt, consider one of thousands of clues the fledgling system grappled with. In the category Diplomatic Relations, it read: "Of the 4 countries the United States does not have diplomatic relations with, the one that's farthest north."

In the first wave of algorithms to assess the clue was a cluster that specialized in grammar. They diagrammed the sentence, much the way a grade school teacher once did, identifying the nouns, verbs, direct objects, and prepositional phrases. This analysis helped to resolve doubts about specific words. The "United States," in this clue, referred to the country, not the army, the economy, or the Olympic basketball team. Then they pieced together interpretations of the clue. Complicated clues, like this one, might lead to different readings, one more complex, the other simpler, perhaps based solely on words in the text. This duplication was wasteful, but waste was at the heart of the Blue J strategy. Duplicating or quadrupling its effort, or multiplying it by 100, was one way it would compensate for its cognitive shortcomings—and play to its advantage in processing speed. Unlike humans, who instantly understand a question and pursue a single answer, the computer might hedge, launching simultaneous searches for a handful of different possibilities. In this way and many others, Blue J would battle the efficient human mind with spectacular, flamboyant inefficiency. "Massive redundancy" was how Ferrucci described it. Transistors were cheap and plentiful. Blue J would put them to use.

While the machine's grammar-savvy algorithms were dissecting the clue, one of them searched for its LAT. In this clue about diplomacy, "the one" evidently referred to a country. If this was the case, the universe of Blue J's possible answers was reduced to a mere 194, the number of countries in the world. (This, of course, assumed that "country" didn't refer to "Marlboro Country" or "wine country" or "country music." Blue J had to remain flexible, because these types of exception often occurred.)

Once the clue was parsed into a question the machine could understand, the hunt commenced. Each expert algorithm went burrowing through Blue J's trove of data in search of the answer. One algorithm, following instructions developed for decoding the genome, looked to match strings of words in the clue with similar strings elsewhere, maybe in some stored Wikipedia entry or in articles about diplomacy, the United States, or northern climes. One of the linguists worked on rhyming key words in the clue or finding synonyms. Another algorithm used a Google-like approach and focused on documents that matched the greatest number of key words in the clue, giving special attention to the ones that surfaced the most often.

While they worked, software within Blue J would compare the clue to thousands of others it had encountered. What kind was it—a puzzle, a limerick, a historical factoid? Blue J was learning to recognize more than fifty types of questions, and it was constructing the statistical record of each algorithm for each type of question. This would guide it in evaluating the results when they came back. If the clue turned out to be an anagram, for example, the algorithm that rearranged the letters of words or phrases would be the most trusted source. But that same algorithm would produce gibberish for most other clues.

What kind of clue was this one on Diplomatic Relations? It appeared to require two independent analyses. First, the computer had to come up with the four countries with which the United States had no diplomatic ties. Then it had to figure out which of them was the farthest north. A group of Blue J's programmers had recently developed an algorithm focused on these so-called nested clues, in which one answer lay inside another. This may sound obscure, but humans ask this type of question all the time. If someone wonders about "cheap pizza joints close to campus," the person answering has to carry out two mental searches, one for cheap pizza joints and another for those nearby. Blue J's "nested decomposition" led the computer through a similar process. It broke the clues into two questions, pursued two hunts for answers, and then pieced them together. The new algorithm was proving useful in *Jeopardy*. One or two of these combination questions came up in nearly every game. They were especially common in the all-important Final Jeopardy, which usually featured more complex clues.

It took Blue J almost an hour for its algorithms to churn through the data and return with their candidate answers. Most were garbage. There were failed anagrams of country names and laughable attempts to rhyme "north" and "diplomatic." Some suggested the names of documents or titles of articles that had strings of the same words. But the nested algorithm followed the right approach. It found the four countries on the outs with the United States (Bhutan, Cuba, Iran, and North Korea), checked their geographical coordinates, and came up with the answer: "What is North Korea?"

At this point, Blue J had the right answer. It had passed the binary recall test. But it did not yet know that North Korea was correct, nor that it even merited enough confidence for a bet. For this, it needed loads of additional analysis. Since the candidate answer came from an algorithm with a strong record on nested clues, it started out with higher than average confidence in that answer. The machine proceeded to check how many of the answers matched the question type "country." After ascertaining that North Korea appeared to be a country, confidence in "What is North Korea?" increased. For a further test, it placed "North Korea" into a simple sentence generated from the clue: "North Korea has no diplomatic relations with the United States." Then it would see if similar sentences showed up in its data trove. If so, confidence climbed higher.

In the end, it chose North Korea as the answer to bet on. In a real game, Blue J would have hit the buzzer. But being a student, it simply moved on to the next test.

The summer of 2007 turned into fall. Real estate prices edged down in hot spots like Las Vegas and San Diego, signaling the end of a housing boom. Senators Obama and Clinton seemed to campaign endlessly in Iowa. The Red Sox marched toward their second World Series crown of the decade, and Blue J grew smarter.

But Ferrucci noted a disturbing trend among his own team: It was slowing down. When technical issues came up, they often required eight or ten busy people to solve them. If a critical algorithm person or a member of the hardware team was missing, the others had to wait a day or two, or three, by which point someone else was out of pocket. Ferrucci worried. Even though the holidays were still a few months away and they had all of 2008 to keep working, his boss, a manager named Arthur Ciccolo, never tired of telling him that the clock was ticking. It was, and Ferrucci—thinking very much like a computer engineer—viewed his own team as an inefficient system, one plagued with low bandwidth and high latency. As team members booked meeting rooms and left phone messages, vital information was marooned for days at a time, even weeks, in their own heads.

Computer architects faced with bandwidth and latency issues often place their machines in tight clusters. This reduces the distance that information has to travel and speeds up computation. Ferrucci decided to take the same approach with his team. He would cluster them. He found an empty lab at Hawthorne and invited his people to work there. He called it the War Room.

At first it looked more like a closet, an increasingly cluttered one. The single oval table in the room was missing legs. So the researchers piggybacked it on smaller tables. It had a tilt and a persistent wobble, no matter how many scraps of cardboard they jammed under the legs. There weren't enough chairs, so they brought in a few from the nearby cafeteria. Attendance in the War Room was not mandatory but an initial crew, recognizing the same bandwidth problems, took to it right away. With time, others who stayed in their offices started to feel out of the loop. They fetched chairs and started working at the same oval table. The War Room was where decisions were being made.

For high-tech professionals, it all seemed terribly old-fashioned. People were standing up, physically, when they had a problem and walking over to colleagues or, if they were close enough, rolling over on their chairs. Nonetheless, the pace of their work quickened. It was not only the good ideas that were traveling faster; bad ones were, too. This was a hidden benefit of higher bandwidth. With more information flowing, people could steer colleagues away from the dead ends and time drains they'd already encountered. Latency fell. "Before, it was like we were running in quicksand," said David Gondek, a new Ph.D. from Brown who headed up machine learning. Like many of the others on the team, Gondek started using his old office as a place to keep stuff. It became, in effect, his closet.

It was a few weeks after Ferrucci set up the War Room that the company safety inspector dropped by. He saw monitors propped on books and ethernet cables snaking along the floor. "The table was wobbly. It was a nightmare," Ferrucci said. The inspector told them to clear out. Ferrucci started looking for a bigger room and quickly realized his team members expected cubicles in the larger space. He told them no, he didn't want them to have the "illusion of returning to a private office." He found a much larger room on the third floor. Someone had left a surfboard there. Ferrucci's team propped it at the entrance and sat a tiny toy bird, a bluebird, on top of it. It was the closest specimen they could find to a blue jay.

A war room, of course, was hardly unique to Ferrucci's team. Financial trading floors and newsrooms at big newspapers had been using war rooms for decades. All of these operations involved piecing together networks of information. Each person, ideally, fed the others. But for IBM, the parallel between the *Jeopardy* team and what it was building was particularly striking. The computer had areas of expertise, some in knowledge, others in language. It had an electrical system to transmit information and a cognitive center to interpret it and to make decisions. Each member of Ferrucci's team represented one (or more) of these specialties. In theory, each one could lay claim to a certain patch of transistors in the thinking machine. So in pushing the team into a single room, Ferrucci was optimizing the human brain that was building the electronic one.

By early 2008, Blue J's scores were rising. On the Jennings Arc posted on the wall of the War Room, it was climbing toward the champion—but was still 30 percent behind him. If it continued the pace of the last six months, it might reach Jennings by mid-2008 or even earlier. But that wasn't the way things worked. Early on, Ferrucci said, the team had taught Blue J the easy lessons. "In those first months, we could put in a new algorithm and see its performance jump by two or three percent," he said. But with the easy fixes in, the advances would be smaller, measured in tenths of a percentage.

The answer was to focus on Blue J's mistakes. Each one pointed to a gap in its knowledge or a misunderstanding: something to fix. In that sense, each mistake represented an opportunity. The IBM team, working in 2007 with Eric Nyberg, a computer scientist at Carnegie Mellon, had designed Blue J's architecture for what they called blame detection. The machine monitored each stage of its long and intricate problem-solving process. Every action generated data, lots of it. Analysts could carry out detailed studies of the pathways and performance of algorithms on each question. They could review each document the computer consulted and the conclusions it drew from it. In short, the team could zero in on each decision that led to a mistake and use that information to improve Blue J's performance.

The researchers were swimming in examples of misunderstandings and wrong turns. Blue J, after all, was failing on half of the clues. But which ones represented larger patterns? Fixing those might enhance its analysis in an entire category. One South America clue, for example, appeared to signal a glitch on Blue J's part in analyzing geography—an important category in *Jeopardy*. The clue asked for the country that shared the longest border with Chile. Blue J came back with the wrong answer: What is Bolivia? The correct response (What is Argentina?) was its second choice.

Analyzing the clue, researchers saw that Blue J had received conflicting answers from two algorithms. The one specializing in geography had come back with the right answer, Argentina, whose 5,308-kilometer border with Chile dwarfed the 861-kilometer Chilean-Bolivian frontier. But another algorithm had counted references to these countries and their borders and found a lot more talk about the Bolivian stretch. (Chile and Bolivia have been engaged in a border dispute since the 1870s, generating a steady stream of news coverage.) Lacking any other context, this single-minded algorithm suggested Bolivia—and Blue J unwisely trusted it. "The computer was paying more attention to popularity than geography," Ferrucci said. Researchers went on to tinker with the ratios underlying Blue J's judgment. They instructed it to give more weight to the geography in that type of question and a bit less to popularity. Then they tested the system on a large batch of similar geography clues. Blue J's performance improved. They then ran it on a group of random clues to find out if the adjustment affected Blue J's performance elsewhere, perhaps turning correct answers into mistakes. That happened all too often. But this time the change helped. Blue J's performance inched ahead another tiny fraction of a percent.

The *Jeopardy* clues, nearly all of them from the J! Archive Web site, were the test bed for this stage of Blue J's education. Eric Brown, Ferrucci's top lieutenant, oversaw this cache along with Chu-Carroll. Brown was serious and circumspect. He got his doctorate at the University of Massachusetts and graduated, in 1996, at the dawn of the dot-com boom. Citing family obligations, he turned down a job offer from Infoseek, one of the early search engines. Two years later, the Walt Disney Company paid $430 million for 42 percent of Infoseek, turning many of the early employees—including the one who grabbed the job Brown had been offered—into multimillionaires. "It's a sad story for me," Brown said. "I ran into him a few years later at a conference. He was retired."

From the very beginning, Brown kept tight control of the *Jeopardy* data. He distributed two thousand clues at a time, which the team used to train Blue J. The risk they faced, as in any statistical analysis, was that they'd fine-tune the machine too precisely to each batch of questions. This tendency to model too closely to a training set is known as overfitting, and it's a serious problem.

Anyone who has ever studied a foreign language knows all about it. Students inevitably overfit to the French or Spanish or Mandarin that their teacher speaks. They adjust to her rhythms and syntax and come to associate that single voice with the language itself. A trip to Paris or Beijing often brings a rude awakening. In Blue J's education, each training set was a single teacher. When the computer started to score well on a training set, the researchers would test it on *Jeopardy* clues it had never seen before. This was a blind set of data, a few thousand clues that no one but Brown had seen. Each time Blue J ventured from its comfortable clues into an unfamiliar set of data, its results would drop about 5 percent. But still, its overall scores were rising. Brown would release another training set, and the process would start over.

The broader question, naturally, was whether the *Jeopardy* challenge itself was one giant exercise in overfitting. *Jeopardy*, in a sense, was a single training set of 185,000 clues, including general knowledge and a mix (that Ferrucci's team quickly quantified) of puzzles, riddles, and the like. If Blue J eventually mastered the game and even defeated Ken Jennings and Brad Rutter in its televised showdown, would its expertise be too specific, or esoteric, for the broader world of business? Would it be flummoxed once it ventured outside its familiar grid of thirty clues? After all, *Jeopardy* champions were hardly famous for running corporations, mastering global diplomacy, or even managing large research projects. They tended to be everyday people—real estate agents, teachers, software developers, librarians—all with one section of their mind specially adapted—or possibly overfitted—to a TV quiz show.

David Ferrucci spent his days swimming in statistics. They defined every aspect of the *Jeopardy* project. Blue J's analysis of data was statistical. Its confidence algorithms and learning programs were fed entirely by statistics. Its choice of words and its game strategy were guided by similar analysis, all statistical. Blue J's climb up the Jennings Arc was a curve defined by statistics, and when it got into sparring sessions with humans, sometime in 2009, its record would be calculated the same way. The Final Match was the rare exception—a fact that haunted Ferrucci from the very start. Blue J's fortunes would be defined more by chance than probability. One game, after all, was a minuscule test set, statistically meaningless. A bit of bad luck on a couple of Daily Doubles, and Blue J could lose—even if statistics demonstrated that it *usually* won.

Ferrucci was constantly analyzing the statistical methodology of teaching and testing the bionic player. One day in the spring of 2008, he came up with a question no one had asked before. Was there any variation, from year to year, in the *Jeopardy* clues? He asked Eric Brown, the guardian of Blue J's training set. Did Blue J fare better against the clues from some years than others?

It was odd, looking back, that such a simple question had gone unasked for so long. It could be important. Even a change in the clue writers or new directions from the producer could usher in new styles or subject matter. By opening up the format to more popular culture in 1997, Harry Friedman had already demonstrated that *Jeopardy*, unlike chess, was a game that changed with the times. Did it evolve in a predictable way? If so, Blue J had to be ready.

Brown's team proceeded to analyze Blue J's performance against the clues, year by year. They were stunned to see that the machine's scores plummeted when answering clues from 2003 and remained at that lower level. It was as if the machine got dumber, by about 10 percent. As Blue J answered the newer questions, its precision stayed constant. In other words, it didn't make more mistakes. But with a lower level of confidence, it didn't buzz as often. Blue J was more confused.

The IBM team called this shift "climate change." For weeks, researchers pored over *Jeopardy* data, trying to figure out why in season 20, from September of 2003 to the following July, the questions suddenly became harder for Blue J. Was it more puzzles or puns? They couldn't tell.

That twentieth season was the one in which Ken Jennings began his remarkable run. Was *Jeopardy* toughening the clues for Jennings and unwittingly making the game harder for Blue J? That seemed unlikely, especially since it would be difficult to make the game harder for the omniscient Jennings without also ratcheting it up for his competitors. Ferrucci and his team asked Friedman about the change. He said he didn't know — and added that at this point IBM certainly knew more about *Jeopardy* clues than he did.

Climate change meant that as Blue J prepared for its first matches with human *Jeopardy* champs — so-called sparring sessions — two-thirds of its training set was too easy. It was like a student who crams for twelfth-grade finals only to see, late in the game, that he's been consulting eleventh-grade textbooks. From Blue J's perspective, the game had just gotten considerably harder.

5

Watson's Face

IN THE FALL of 1992, a young painter named Joshua Davis moved from Colorado to New York City and enrolled at the prestigious Pratt Institute. After a year, he switched from painting to illustration, where there were better career opportunities. "I thought, 'I'll still paint. It'll just be for the Man,'" Davis said. But when he sent his work to two book publishers, hoping to line up illustration contracts for children's books, the response was essentially, as he put it, "'Thanks but no thanks, and like, who the fuck are you?'"

Davis didn't take it too hard. His self-esteem was strong enough to withstand a knock or two. A bit later a friend at school steered him toward the digital world. "He said, 'Oh, don't worry, man, there's this whole Internet thing now. Like books are dead.'" Davis said he was "totally naive" at that point. "I said, 'Cool. Print's dead. Fantastic!'" He promptly bought an old computer, but it lacked an operating system. So he went to a bookstore and bought one last artifact from the printed world: a manual for the new open-source system called Linux. A diskette he found at the back of the book contained the software. "I was like, 'Score!'" he said.

Davis didn't know he was about to tackle what he calls the "world's hardest operating system." But as he taught himself about user interface design, programming, and video graphics, he had an epiphany. He wasn't going to use computers simply to create designs more quickly or to reach more people. The technology itself, following his instructions, would generate the art. "At the time I thought, 'The Internet is my new canvas,'" he said.

His first corporate job was for Microsoft. He designed visual applications for Internet Explorer 4, which debuted in 1997. For the next few years, he became a leader in the new field of generative art, using programs to combine data into colors and patterns that could morph into countless variations. For this he harnessed movements from nature, such as wind, flowing water, and swarming birds and insects. He even turned his body into an evolving canvas. He had his entire left arm tattooed with the twenty glyphs of the Mayan calendar, the swirling designs running up his right arm depicted Japanese wind, and his back carried images of water. Fire, he said, would eventually cover his chest. He had birds tattooed on his neck, one of them dedicated to his daughter, Kelly Ann.

Davis built a thriving studio, with offices in Manhattan and Chicago, and a long list of clients, from Volkswagen and Motorola to rap luminaries Sean "Puff Daddy" Combs and Kanye West. He eventually moved from the city to a hundred-year-old house with a barn in Mineola, on Long Island. As his success grew, he gave more thought to where his work fit in the history of art. In 2008, for a lecture series on dynamic abstraction, he focused on Jackson Pollock, the abstract artist famous for dripping paint on canvases from a stepladder. "Here's a guy who says, 'I'm going to paint, but I'm going to use gesture.'" Davis waved his arms to illustrate the movement. "Wherever the paint goes, the paint goes." Not one to sell himself short, he said he felt like an extension of Pollock. "I'm creating systems where I establish the paints, the boundaries, and the colors. But where it goes is where it goes. It's like controlled chaos."

As Davis learned more about Pollock, his feelings of kinship only grew. He read that the other artist had also left the city, moved to Long Island, and worked in a barn. "It was, like, sweet!" Davis said. "How did *that* work out?"

It was around that time, in October 2008, that Davis got a call from an art director at Ogilvy & Mather, the international advertising agency. IBM, he learned, was building a computer to take on human champions in *Jeopardy*. How would he like to create the machine's face?

During the first year of Blue J's development, few at IBM thought much about the computer's physical presence or its branding. A pretty face would be irrelevant if the team couldn't come up with a workable brain. But by late summer of 2008, Ferrucci's team was getting close. One August day, Harry Friedman and the show's supervising producer, a former *Jeopardy* champion named Rocky Schmidt, visited the Yorktown labs for their first look at the bionic player.

As the group gathered in one of the windowless conference rooms at the Yorktown lab, Ferrucci walked them through the computer's cognitive process, explaining how it came up with answers and why, on occasion, it flubbed them so badly. He explained that the hardware—what would become Watson's body—wasn't yet ready to deliver timely answers. But the team had led the computer through a game of *Jeopardy*, had recorded its answers, and then created a simulation of the game by loading the answers into a laptop. With that, Friedman and Schmidt watched the new contestant in action. Friedman later said that he had been "blown away" by the computer's performance.

The conversation, according to Noah Syken, a media manager at IBM, quickly turned to logistics and branding. If the computer required the equivalent of a roaring data center to play the game, where would all that machinery fit on the *Jeopardy* set? And how about all the noise and heat it would generate? One possibility might be to set up its hulking body on the *Wheel of Fortune* set, next door, and run the answers to the podium. But that raised a bigger question: What would viewers see at that podium? No one had a clue.

The following month, as Lehman Brothers imploded, car companies crashed, and the world's financial system appeared to teeter on the verge of collapse, IBM's branding and marketing team worked to develop the personality and message of the *Jeopardy*-playing machine. It would need a face of some sort and a voice. And it had to have a name.

An entire corporate identity unit at IBM specialized in naming products and services. A generation earlier, when the company still sold machines to consumers, some of the names this division dreamed up became iconic. "PC" quickly became a broad term for personal computers (at least those that weren't made by Apple). ThinkPad was the marquee brand for top-of-the-line business laptops. And for a few decades before the PC, the Selectric, the electric typewriter with a single rotating type ball (which could "erase" typos with space-age precision) epitomized quality for anyone creating documents. With IBM's turn toward services, the company risked losing its contact with the popular mind—and its identity as a hotbed of innovation.

What's more, a big switch had occurred since the 1990s.

It used to be that the most advanced machinery was found at work. Children whose parents went to offices would sometimes get a chance to play with the adding machines there, along with the intercoms, fancy photocopiers, and phones with illuminated buttons for five or six different lines. But at the dawn of the new century, the office appeared to lose its grip on cool technology. Now people often had snazzier gadgets at home, and in their pockets, than at work. Companies like Apple and Google targeted consumers and infused technology with fun, zip, even desire. Tech companies that served the business market, by contrast—Oracle, Germany's SAP, Cisco, and IBM—tended to stress the boring stuff: reliability, efficiency, and security. They were valuable qualities, to be sure, but deadening for a brand. IBM needed some sizzle. It was competing for both investors and brainpower with the likes of Google, Apple, Facebook—even the movie studio Pixar. It had to establish itself in the popular imagination as a company that took risks and was engaged in changing the world with bleeding-edge technology. The *Jeopardy* challenge, with this talking IBM machine on national television matching wits with game-show luminaries, was the branding opportunity of the decade. The name had to be good.

Was THINQ the right choice, or perhaps THINQER? How about Exaqt or Ace? Working with the New York branding firm VSA Partners, IBM came up with dozens of candidates. The goals, according to a VSA summary, were to emphasize the business value of the technology, create a close tie to IBM, steer clear of names that were "too cute," and lead the audience "to root for the machine."

One group of names had strong links to IBM. Deep Logic evoked Deep Blue, the computer that mastered chess. System/QA recalled the iconic mainframe System/360. Other names stressed intelligence. Qwiz, for example, blended "Q," for question, with "wiz" to suggest that the technology had revolutionized search. The pronunciation—quiz—fit the game show theme. Another choice, nSight, referred to "n," representing infinite possibilities. And EureQA blended "eureka" with the Q-A for question-answering. Another candidate "Mined," pointed to the machine's datamining prowess.

On the day of the naming meeting, December 12, all of the logic behind the various choices was promptly ignored as people focused on the simplest of names in the category associated with IBM's brand: Watson. "It just felt so right," said Syken. "As soon as it came up, we knew we had it." Watson invoked IBM's founder. This was especially fitting since Thomas J. Watson had also established the research division, originally on the campus of Columbia University, in 1945. The Watson name was also a nod to the companion and chronicler of Sherlock Holmes, the brilliant fictional sleuth. In those stories, of course, Dr. Watson was clearly the lesser of the two intellects. But considering public apprehension about all-knowing machines, maybe it wasn't such a bad idea to name a question-answering computer after an earnest and plodding assistant.

The next issue was what Watson would look like. For this, IBM brought in its lead advertising agency, Ogilvy & Mather. With offices on Manhattan's sprawling far West Side, where it shared a block with a Toyota dealership and a car wash, Ogilvy had been IBM's primary agency since Louis Gerstner arrived at the company. Its creative minds were paid to think big, and in the first few meetings, they did. They considered creating an enormous wall of Watson. It would take over much of the *Jeopardy* set, perhaps in the form of a projected brain, with neurons firing, or maybe a virtual sketchpad, dancing with algorithms and formulas as the machine cogitated. "They were pretty grand ideas," said David Korchin, the project's creative director.

In talking to *Jeopardy* executives, though, it quickly became clear that they'd have to think smaller. If IBM's Watson passed muster, it would be a guest on the show. It would not take it over, Its branding space, like that of any other contestant, would be limited to the face behind the podium—or whatever fit there. *Jeopardy* held the power and exercised it. If IBM's computer was to benefit from an appearance on *Jeopardy*, the quiz show would lay down the rules.

Now that Watson was reduced from a possible Jumbotron to a human-sized space, what sort of creature would occupy it? "Would it look like a human?" asked Miles Gilbert, the art director. "Would it be an actual human? Was there a single person who could represent IBM?" At one point, he said, they considered establishing Watson as a child, one that learns and grows through an educational process. That didn't make sense, though, because Watson would already be an adult by the time it showed up on TV. (And *Jeopardy* apparently wasn't going to give IBM airtime to describe the education of young Watson.) The Ogilvy team also considered other types of figures. A new Pixar movie that year featured Wall-E, a lovable robot. Perhaps that was the right path for Watson.

Whether it was a cartoon figure or a bot like Wall-E, much of the discussion boiled down to how human Watson should be. The marketers feared that millions of viewers might find it unsettling if the computer looked or acted too much like a real person. Science fiction was full of evil "human" computers. HAL, the mutinous machine running the spaceship in Stanley Kubrick's *2001: A Space Odyssey*, was the archetype. It killed four of the five astronauts on board. The last one had to remove the machine's cognitive components one by one to save his own life. "We didn't need this project and Watson to scare people about technology," said Syken. "If you go to our YouTube channel and see the comments, you'll see people talking about *2001* again and again, and IBM tracking people." He had a point. In one short IBM video about technology in neonatal care, someone with the username Present1os commented: "This is creepy. Reminds me of 'Invasion of the Body Snatchers.' Also a multinational taking over human bodies."

Another thorny issue for IBM was jobs. Big Blue, perhaps as much as any company, was known for replacing people with machines. That was the nature of technology. In the 1940s, IBM turned its attention to the world's industrial supply chains, the enormously complex processes that wound their way from the loading docks of iron mines to the shiny bumpers in a Cadillac showroom, from cattle herds in Kansas to the vendor selling hot dogs in Yankee Stadium. Each of these chains wound its way through depots, rolling mills, slaughterhouses, and packaging plants, providing jobs at every step. But these processes had evolved willy-nilly over the years and weren't efficient. By building mathematical models of the supply chains, IBM could help companies cut out waste and duplication, speeding them up and slashing costs. This process, known as optimization, often eliminated jobs. The engine of optimization, and its symbol, was a big blue IBM mainframe computer.

In the following decades, computers continued to replace people, supplanting bank tellers, toll collectors, and night watchmen. Steel mills as big as cathedrals, which once crawled with workers, operated with skeleton crews, most of them just monitoring the computerized machinery. Robots moved on to automobile assembly lines. Good arguments could be made, of course, that inefficient companies faced extinction in a competitive global economy. In that sense, optimization and automation *saved* jobs. And in a healthy economy, workers would migrate toward more productive sectors, even if the transition was often painful. The quickly growing tech industry itself employed millions. For many, though, textbook economics and distant success stories provided little comfort. Computers, in the popular mind, killed jobs.

And IBM was producing ever more sophisticated models. Researchers in the company labs in Yorktown, New York, and Cambridge, Massachusetts, were applying many of the lessons learned in the industrial world to the modern workplace. With a computerized workforce, like IBM's own, each employee left an electronic record of clicks, updates, e-mails, and jobs completed. Researchers could analyze individual workers—their skills, the jobs they did, the effectiveness of their teams. The goal was to fine-tune the workforce. "We evaluate every job," said Samer Takriti, who headed a study of company workers at IBM Research until 2007, "and we calculate whether it could be handled more efficiently offshore or by a machine."

Given this type of analysis, it wasn't hard to imagine that millions of television viewers might regard a question-answering computer as a fearsome competitor rather than a technological marvel. What's more, as the IBM team discussed these issues in the fall of 2008, the global economy seemed to be collapsing. Watson's turn on *Jeopardy* might well take place during a period of growing joblessness and economic fear. It could be the next Great Depression. In such a climate, they decided, a humanoid Watson might frighten people. In response, they moved to focus the publicity campaign less on the machine than on the team that built it. "This had to be a story about people," said Syken.

When it came to Watson's avatar, IBM and Ogilvy chose to avoid anything that might make it look human, opting for abstraction. The outlines of this avatar, as it turned out, were already taking shape in another division of the company. For a year, IBM's global strategy team had been developing a campaign to communicate Big Blue's technologies, and its mission, in a simple slogan. In a company with four hundred thousand people and hundreds of business lines, this was no easy task. What they settled on was data. In the modern economy, nearly every machine received instructions from the computer chips inside it. Many were already linked to networks, and others soon would be. These machines produced ever-growing rivers of digital data that detailed, minute by minute, the operations taking place across the planet. Many of these processes, such as bus routes, hospital deliveries, the patterns of traffic lights, had simply evolved over time, like the old industrial supply chains. They seemed to work. But given the data and much more that was en route, mathematically savvy analysts were able to revamp haphazard systems, saving time and energy. Science would replace intuition. The electrical grids, infused with new information technology, would grow smarter, predicting demand—house by house, business by business—and providing just the right amount of current to each user. Traffic patterns would be organized to reduce congestion and pollution. So would garbage collection, the delivery of health care and clean water, and the shuttling of farm goods to the cities. These intelligent systems were IBM's niche. Technology would lead to what the company called a Smarter Planet.

Two days after the election of Barack Obama as president,

on November 6, 2008, IBM's chairman, Sam Palmisano, appeared before the Council on Foreign Relations in New York City to unveil the Smarter Planet initiative. He framed it in the context of the global economic crisis, saying that the world would adopt these approaches "because we must." He said that carrying on with the status quo, running business and government the traditional twentieth-century way, had led to the economic and environmental crises and was "not smart enough to be sustainable." Illustrating his talk was an icon of the planet Earth with five bars radiating from it. This was Chubby Planet, and the bars represented intelligence.

Chubby Planet soon became the basis of Watson's avatar. It made sense. Chubby was abstract. It represented intelligence. And it fit into IBM's global branding effort. In one form or another, the Watson version of Chubby Planet would express the machine's cognitive processes—without betraying emotion. The IBM-Ogilvy team decided that the computer would answer the questions in a friendly, even-keeled male voice. It would not change with the flow of the game. No voiced frustration, no regrets, and certainly no gloating.

But at the heart of the decision-making process was a paradox: A company built on scientific analysis was running a global branding campaign from intuition. Before launching any new product, IBM had the means and expertise to carry out sophisticated tests analyzing public reaction. The research division had an entire social media unit, in Cambridge, Massachusetts, that specialized in new methods of tracking consumer sentiments through the shifting words and memes cascading across the Internet and sites like Twitter and Facebook. IBM's consultants around the world were helping other companies tie these studies to their businesses. Yet when it came to creating the face, voice, and personality of its own game-

playing computer, the IBM team relied on instincts—a vague
sense they had of consumers' interests and fears. IBM and
Ogilvy ran the campaign in a way that Watson could never
compute: from the gut.

This isn't to say that statistical analysis would have pointed
IBM toward an ideal form and personality for Watson. People's attitudes about computers, and what they should express, were complicated, and they varied—by generation, geography, and gender. Clifford Nass, a Stanford professor and
author of *The Man Who Lied to His Laptop,* studies the relations between humans and their machines. In one experiment, people played blackjack against a computer. The computer was represented by a photograph of a person along with
a cartoon-like bubble for text. In one scenario, the computer
expressed interest only in itself—"I'm happy, I won." In another, it empathized only with its opponent, and in a third,
it expressed feelings for both itself and its opponent. The humans in the test certainly didn't like the self-centered computer. But the males in the test group preferred it when the
computer showed interest only in them, while females favored
the balanced approach.

One lesson from this and other studies, according to Nass,
is that people quickly develop feelings, from admiration to
resentment, for the machines they encounter. And this was
sure to be the case for millions when they saw Watson playing
Jeopardy on their television. He argued that people would feel
more positively toward a computer that expressed feelings to
match its performance. "That computer had better have some
emotion," he said. "It should sound stressed if it's not doing
well." If it didn't express emotions, he said, it would seem
alien and perhaps menacing. "When it sits there and it's not
clear what it wants, we think, 'What the hell is going on?'"

he said. "The scariest movies are when you don't know what something wants."

In IBM's defense, even if the company had wanted to provide Watson with a rich and modulated human voice, it would have required a large development effort to build it. Existing voice technology came close to expressing human emotion but was still a bit off. The IBM team worried that people would resent, or fear, a computer that tried to mimic the emotional range of the human voice and fell short. To save money and reduce that risk, they adapted a friendly bionic voice they already had on a shelf. This Watson would remain relentlessly upbeat through the ups and downs of its *Jeopardy* career.

Not that the avatar wouldn't be expressive in its own way. Working in his Long Island studio, Joshua Davis was devising schemes to represent Watson's cognitive processes. He worked with forty-two threads of color that would stream and swarm across Watson's round avatar according to what was going on in its "mind." It would look a bit like the glass globes at science museums, where each touch of a human hand summons jagged ribbons of lightning. Davis, a sci-fi buff, picked the number forty-two as an homage to Douglas Adams's *Hitchhiker's Guide to the Galaxy*. In that book, a computer named Deep Thought calculated the answer to the Ultimate Question of Life, the Universe, and Everything. It was forty-two. Someday, perhaps, a smarter computer would come up with the question for which forty-two was the answer. For Davis, the forty-two threads were his own little flourish within the larger work. "It's my Easter egg," he said.

But what stories would those threads be telling? The Ogilvy team started by dissecting videos of *Jeopardy* games. They divided the game into the various states. They began with the booming voice of the longtime announcer, Johnny Gilbert, saying, "*This* is *Jeopardy!*" They continued through the applause, the introduction of the contestants and the host, Alex Trebek, and every possible permutation after that: when Watson won the buzz, when it lost, when the other player chose a category, and when the contestants puzzled over Final Jeopardy, scribbling their responses. There were a total of thirty-six states, each with its prescribed camera shot, many of them just a second or two long. (Davis was disappointed that they couldn't find six more, raising it to his magical number. "If we could just get it to forty-two," he joked, "I'm pretty sure something quantum mechanical could happen, like a tornado of butterflies.")

Still, it was clear that unless Watson got special treatment, the avatar would garner precious little screen time. When it answered a question, the camera would be focused on it for between 1.7 and 5 seconds. And during its most intense cognitive stages—when it was considering a question, going through the millions of documents, and choosing among candidate answers—the camera would stay fixed for a crucial 3 or 4 seconds on the clue. In essence, Davis had to prepare an avatar for a series of cameo appearances. He said he was unfazed. "Watson is that ultimate challenge," he said. "I've got milliseconds of time where I need to present something that's compelling and dynamic." He went about developing different patterns for the thirty-six cognitive states in the computer. The threads would flow into a plethora of patterns and colors as it waited, listened, searched, buzzed, and pronounced its answer. The threads would soar when Watson bounded with confidence, droop when it felt confused.

While all of this work was in progress, the *Jeopardy* challenge remained a closely guarded secret. But that changed in the spring of 2009. IBM's top executives, excited about the prospect of the upcoming match, wanted to highlight it in the company's annual shareholder meeting, scheduled for April 28 at the Miami Beach Convention Center. To prepare for the media coverage sure to follow, the computer scientists on Ferrucci's team were ferried into New York City to receive media training. They were instructed to focus on the human aspect of their venture—the people creating the machine—and to avoid broader questions concerning IBM, such as the company's financial prospects or its growing offshore business.

Only one problem. The agreement IBM and *Jeopardy* had in place was little more than a handshake. They had to nail it down. IBM, said executives, was hoping to hammer out a deal that would include airtime for corporate messaging, perhaps telling the history of Watson, how it worked, and what such machines portended for the Information Age. But once again, Harry Friedman and his *Jeopardy* colleagues had all the leverage. IBM needed an agreement right away. *Jeopardy* did not. So Big Blue got a tentative deal, pending Watson's performance over the coming year, in time for the Miami meeting. But other than that, the negotiators came back from Culver City empty-handed, with no promises of extra airtime or other promotional concessions.

Not everything hinged on the final game. IBM hoped that Watson would enjoy a career long after the *Jeopardy* showdown. They had plans for it to tour extensively, perhaps at company events or schools. This mobile Watson might be just a simulation, running on a laptop. Or maybe they could run the big Watson, the hundreds or thousands of processors at the Hawthorne labs, from a remote pickup. The touring Watson would have advantages, at least from Joshua Davis's perspective. Freed from the constraints of *Jeopardy* production, people would have more time to study the changing moods and states of Watson's avatar. Of course, even the touring machine would have to comply with the provisions surrounding *Jeopardy*'s brand and programming. That meant more negotiations, most likely with Harry Friedman still holding most of the cards.

Even as the avatar took shape, no one knew what sort of display it would run on. Davis and the Ogilvy team considered many options to house the avatar, including one technology that projected holograms on a pillar of fog. But they eventually turned to more traditional displays. In that realm, few could compete with Sony, *Jeopardy*'s parent company. Friedman said that Sony engineers conceivably could create a display for Watson, but that such an effort would probably require a call from IBM's Sam Palmisano to Sony's top executive, Howard Stringer. "We said, 'That's not going to happen,'" said one IBM executive. "We'll save that call for something more important." Still Sony had a possibility. In December, a team of five Sony employees flew from Tokyo to the Yorktown labs with a prototype of a new display. It was a projection technology so secret, they said, that no one could even take pictures of it. IBM considered it a bit too small for the Watson avatar, the Japanese contingent flew home, and the search continued.

Vannevar Bush, the visionary who in the 1940s imagined a mechanical World Wide Web, once wrote that "electronic brains" would have to be as big as the Empire State Building and require Niagara Falls to cool them. Of course, the computers he knew filled entire rooms, were built of vacuum tubes, and lacked the processing power of a hand-me-down cell phone. While Davis continued to develop Watson's face,

Ferrucci's team started to grapple with a new challenge. To date, their work had focused on building software to master *Jeopardy*. Watson was only a program, like Microsoft's Windows operating system or the video game Grand Theft Auto. To compete against humans, the Watson program would have to run on a specially designed machine. This would be Watson's body. It might not end up as big as a skyscraper, but it would be a monster all the same. That much was clear.

The issue was speed. The millions of calculations for each question exacted a price in time. Millisecond by millisecond, they added up. Each clue took a single server an average of 90 minutes to 2 hours, more than long enough for Jennifer Chu-Carroll's lunch break. For Watson to compete in *Jeopardy*, Ferrucci's team had to shave that down to a maximum of 5 seconds and an average of 3.

How could Watson speed up by a factor of 1,440? In late 2008, Ferrucci entrusted this job to a five-person team of hardware experts led by Eddie Epstein, a senior researcher. For them, the challenge was to divide the work Watson carried out in two hours into thousands of stand-alone jobs, many of them small sequences of their own. They then had to distribute each job to a different processor for a second or two before analyzing the results cascading in. This work, or scale-out, required precise choreography—thousands of jobs calculated to the millisecond—and it would function only on a big load of hardware.

Epstein and his team designed a chunky body for Watson. Packing the computers closely limited the distance the information would have to travel and enhanced the system's speed. It would develop into a cube of nearly 280 computers, or nodes, each with eight processors—the equivalent of 2,240 computers. The eight towers, each the size of a restaurant refrigerator, carried scores of computers on horizontal shelves, each about as big as a pizza box. The towers were tilted, like the one in Pisa, giving them more surface area for cooling. In its resting state, this assembly of machines emitted a low, whirring hum. But about a half hour before answering a *Jeopardy* question, the computers would stir into action, and the hum would amplify to a roar. During this process, Watson was moving its trove of data from hard drives onto random access memory (RAM). This is the much faster (and more expensive) memory that can be searched in an instant—without the rotating of disks. Watson, in effect, was shifting its knowledge from its inner recesses closer to the tip of its tongue. As it did, the roar heightened, the heat mounted, and a powerful set of air conditioners kicked into high gear.

It was a snowy day in February 2010 when the marketing team unveiled prototypes of the Watson avatar for David Ferrucci. Ferrucci was working from home with a slow computer connection, so it took him several long minutes to download the video of the avatar in action. "It's amazing we can get a computer to answer a question in three seconds and it still takes fifteen minutes to download a file," he muttered. When he finally had the video, the creative team walked him through different possible versions of Watson. They weren't sure yet whether the avatar would reside in a clear globe, a reddish sphere, or perhaps a simple black screen. However it was deployed, it would continuously shift into numerous states of listening and answering. Miles Gilbert, the art director, explained that the five bars of the Smarter Planet icon would stay idle in the background "and then pop up when he becomes active."

"This is mesmerizing," Ferrucci said. But he had some

complaints. He thought that the avatar could show more of the computation going on inside the machine. Already, the threads seemed to simulate a cognitive process. They came from different parts of the globe and some grew brighter while others faded. This was actually what was happening computationally, he said, as Watson entertained hundreds of candidate answers and sifted them down to a handful and then just one. Wouldn't it be possible to add this type of real-time data to the machine? "It would be neat if all this movement was less random and meant more," he said.

It sounded like an awful lot of work for something that might fill a combined six minutes of television time. "You're suggesting that there should be thousands of threads, and then they're boiled down to five threads, and ultimately one?" asked a member of the research division's press team.

"Yeah," Ferrucci said. "These are threads in massive parallelism. As they come more and more together, they compete with each other. Then you're down to the five we put on the [answer] panel. One of them's the brightest, which we put into our answer. This," he said emphatically, "could be more precise in its meaning."

There was silence on the line as the artists and PR people digested this contribution from the world of engineering. They moved on to the types of data that Watson could produce for its avatar. Could the system deliver the precise number of candidate answers? Could it show its levels of confidence in each one rising and falling? Ferrucci explained that the machine's ability to produce data was nearly limitless—though he wanted to make sure that this side job didn't interfere with its *Jeopardy* play. "I'm tempted to say something I'll probably regret," he said. "We can tell you after each question the probability that we're going to win the game." He laughed. "Is there room for that analysis?"

It was around this time that Ferrucci, focusing on the red circular version of Watson, started to carry out image searches on the Internet. He was looking for Kubrick's *2001*. "You probably want to avoid that red-eye look," he said, "because when it's pulsating, it looks like HAL. I'm looking at the HAL eye on the Web. It's red and circular, and kind of global. It's sort of like Smarter Planet, actually."

The call ended with Ferrucci promising new streams of Watson data for Joshua Davis and his colleagues at Ogilvy. They had at least until summer to get the avatar up and running. But the rest of Watson—the faceless brain with its new body—was heading into its first round of sparring matches. They would be the first games against real *Jeopardy* players, a true test of Watson's speed, judgment, and betting strategy. The humans would carry back a trophy, along with serious bragging rights, if they managed to beat Watson before Ken Jennings and Brad Rutter even reached the podium.

6

Watson Takes On Humans

EARLY IN THE MATCH, David Ferrucci sensed that something was amiss. He was in the observation room next to the improvised *Jeopardy* studio at IBM Research on a midwinter morning in 2010. On the other side of the window, Watson was battling two humans—and appeared to be melting under the pressure. One Daily Double should have been an easy factoid: "This longest Italian river is fed by 141 tributaries." Yet the computer inexplicably came up with "What is _____?" No Tiber, no Rubicon, no Po (the correct response). It didn't come up with a single body of water, Italian or otherwise. It drew a blank.

Ferrucci leaned forward, looking agitated, and said to no one in particular, "It doesn't feel right. Did you leave off half the system?" His colleagues, all typing on their laptops, kept their heads down and murmured that they hadn't. To engage Ferrucci when he was in a darkening mood could backfire. No one was looking for a confrontation this early in the morning.

Watson continued to malfunction. As the two *Jeopardy* players outscored the machine, it developed a small speech defect. Its genial male voice started to add a "D" to words ending in "N." In the category the Second Largest City, Watson buzzed for the clue, Lahore, and confidently answered, "What is Pakistand?" After a short consultation, the game judge, strictly following the rules, declared the answer incorrect. That turned Watson's $600 gain into a loss, a difference of $1,200. "This is ridiculous," Ferrucci muttered.

Then Watson, a still faceless presence at the far left podium, began to place some ludicrous bets. In one game, it was losing to a journalist and former *Jeopardy* champion named Greg Lindsay, $12,400 to $6,700. Watson landed on a Daily Double. If it bet big, it could pull even with Lindsay or even inch ahead. Yet it wagered a laughable $5. It was Watson's second strange bet in a row. The researchers groaned in unison. Some of their colleagues were sitting in the studio with the *New York Times Magazine*'s Clive Thompson, who was writing a piece on Watson. They looked through the window at Ferrucci and shrugged, as if to ask "What's up with this beast?"

But Ferrucci didn't see them. He was staring at David Gondek. Lithe and unusually cheerful, Gondek was a leading member of the team. Unlike most of his suburban colleagues, he lived far south in Greenpoint, Brooklyn, taking the train and biking from the station. He headed up machine learning and game strategy and seemed to have a hand in practically every aspect of Watson. Ferrucci continued to stare wordlessly at him. Gondek, after all, was responsible for programming Watson's betting strategy, and it looked like the computer was playing to lose. Ferrucci, during this brief interlude, was carrying out an inquisition with his eyes.

Gondek looked up at his boss. "It's a heuristic," he explained. He meant that Watson was placing bets according to a simple formula. Gondek and his colleagues were hard at work on a more sophisticated betting strategy, which they hoped would be ready in a month. But for now, the computer relied on a handful of rules to guide its wagers.

"I didn't realize that it was this stupid!" Ferrucci said. "You never told me it was brain-dead." He gestured toward Thompson, who was watching the game on the other side of the glass and taking notes on his laptop. "We really enjoy stinking it up for the *New York Times* writer."

Gondek started to explain the thinking behind the heuristic. If Watson had barely half the winnings of the leader, one of its rules told it not to risk much in a Daily Double. Its primary goal at this point was not to catch up but to reach Final Jeopardy within striking distance of the leader. If it fell below half of the leader's total, it risked being locked out of Final Jeopardy—a disaster. So Watson was instructed to be timid in these circumstances, even if it meant losing the game—and infuriating the chief scientist.

Nearly every week for several months, IBM had been bringing in groups of six players with game experience to match wits with Watson in this new mock-*Jeopardy* studio. They competed on game boards that had already been played in Culver City but not yet telecast. Friedman's team would not grant IBM access to the elite players who qualified for *Jeopardy*'s Tournament of Champions. They didn't want to give Watson too much exposure to *Jeopardy* greatness—at least not yet. For sparring partners, the machine had to settle for mere mortals, players who had won no more than two games in televised matches. It was up to Ferrucci's team to imagine—or, more likely, to calculate—how much more quickly Ken Jennings and Brad Rutter would respond to the buzzer and how many more answers they'd get right.

By the time Watson started the sparring sessions, in November 2009, the machine had already practiced on tens of thousands of *Jeopardy* clues. But the move from Hawthorne to the Yorktown research center placed the system in a new and surprising laboratory. Playing the game tested new skills, starting with speed. For two years, development at the Hawthorne labs had focused on Watson's cognitive process—coaxing it to come up with right answers more often, to advance up the Jennings Arc. During games, though, nailing the answer meant nothing if Watson lost the buzz. At the same time, it had to grapple with strategy. This meant calculating its bets in Daily Doubles and Final Jeopardy and estimating its chances on clues it had not yet seen. It also had to anticipate the behavior of its human foes, especially in Final Jeopardy, where single bets often won or lost games.

Perhaps the biggest revelation in the sparring matches came from the spectators: They laughed. They were mostly friends of the players and a smattering of IBM employees, watching from four rows of folding chairs. Watson amused them. This isn't to say that they weren't impressed by a machine that came up with some of the most obscure answers in a matter of seconds. But when Watson committed a blooper—and it happened several times a game—they cracked up. They laughed when Watson, exercising its mastery of roman numerals, referred to the civil rights leader Malcolm X as "Malcolm Ten." They laughed more when Watson, asked what the "Al" in Alcoa stood for, promptly linked the aluminum giant to one of America's most notorious gangsters: "What is Al Capone?" (Watson, during this stage, often referred to people as things. This established a strange symmetry, since the contestants routinely referred to the *Jeopardy* machine as "him.") One Final Jeopardy answer a few weeks later produced more merriment. In the category 19th Century Literature, the clue read: "In Chap. 10, the whole mystery of the handkerchiefs, and the watches, and the jewels . . . Rushed upon this title boy's 'mind.'" Instead of Oliver Twist, Watson somehow came up with a British electronic dance music duo, answering, "What is the Pet Shop Boys?"

From a promotional perspective, an occasional nonsensical answer promised to make Watson a more entertaining television performer, as long as the computer kept it clean. This wasn't always assured. In one of its first sparring sessions, in late 2009, the machine was sailing along, thrashing a couple of mid-level *Jeopardy* players in front of an audience that included Harry Friedman and fellow *Jeopardy* bosses. Then Watson startled everyone with a botched answer for a German four-letter word in the category Just Say No. Somehow the machine came up with "What is Fuck?" and flashed the word for all to see on its electronic answer panel. To Watson's credit, it didn't have nearly enough confidence in this response to buzz. (It was a human who correctly responded, "What is *nein*?") Still, Ferrucci was mortified. It was a relief, he said, to look over at Friedman and his colleagues and see them laughing.

Still, such a blunder could tarnish IBM's brand. Watson was the company's ambassador. It was supposed to represent the future of computing. Machines like this, the company hoped, would soon be answering questions in businesses around the world. But it was clear that Watson could conceivably win the *Jeopardy* challenge and still be remembered, on YouTube and late-night TV, for its gaffes. After an analysis of Watson's errors, IBM concluded that 5 percent of them were "embarrassing." This led Ferrucci, early in 2010, to assign a team of researchers to a brand-new task: keeping Watson from looking dumb. "We call it the stupid team," said Chu-Carroll. Another team worked on a profanity filter.

As each day's sparring sessions began, the six *Jeopardy* players settled into folding chairs between the three contestant podiums, the host's stand, and the big *Jeopardy* board, with its familiar grid of thirty clues. David Shepler stood before them. Dark, thin, and impeccably dressed, Shepler ran the logistics of the *Jeopardy* project. He sweated the details. He made sure that IBM followed to the letter the legal agreements covering the play. He didn't bend an inch for Watson. (It was his ruling that docked Watson $600 for mispronouncing Pakistan.) In the War Room's culture of engineers and scientists, Shepler, a former U.S. Air Force intelligence officer, was an outsider. He told them what they could not do, which at times led to resentment. Before each match, he instructed the contestants on the rules. They weren't to tell anyone or—heaven forbid—blog about the matches, the behavior of Watson, or the clues, which had been entrusted to IBM by *Jeopardy*. He had them sign lengthy nondisclosure agreements and then introduced David Ferrucci.

On this winter morning, Ferrucci ambled to the front of the room. He was wearing dark slacks and a black pullover bearing IBM's logo. He outlined the *Jeopardy* challenge and described the goal of building a question-answering dynamo. He pointed to the window behind them, where a set of blue rectangular towers housed the computers running the Watson program. Through the double-pane window, the players could hear the dull roar of the fans working to cool its processors. Ferrucci, priming the humans for the match ahead, tossed out a couple of *Jeopardy* clues, which they handled with ease. "Oh, I bet Watson's getting nervous," he said. "He could be in for a tough day."

Still, Watson had made astounding progress since its early days in the War Room. Ferrucci showed a slide of what used to be the Jennings Arc. It had the same constellation of Jennings dots floating high and to the right. But it had been expanded into a Winners Cloud, with blue dots representing hundreds of other *Jeopardy* winners. Most of the winners occupied the upper right quadrant, but below and to the left of most of Jennings's dots. The average winner buzzed on about half the questions and got four out of five right. Ferrucci traced Watson's path on the chart. The computer, which in 2007 produced subhuman results, now came up with confident answers to about two-thirds of the clues it encountered and got more than 80 percent of them right. This level of performance put it smack in the middle of the Winners Cloud. Though not yet in Ken Jennings's orbit, but it was moving in that direction. Of its thirty-eight games to date against experienced players, Ferrucci said, it had won 66 percent, coming in third only 10 percent of the time.

While explaining Watson's cognitive process, Ferrucci pointed to a black electronic panel. The players wouldn't see it during the game, he explained, but this panel would show the audience Watson's top five candidate answers for each question and how much confidence the machine had in each one. "This gives you a look into Watson's brain," he said. Moments later, he gave them a glimpse into his own. Showing how the computer answered a complicated clue featuring the Portuguese explorer Vasco da Gama, Ferrucci pointed to the list of candidate answers. "I was confident and I got it right," he said. Then, realizing that he was doing a mind meld, he explained that he was speaking for Watson. "I identify with the computer sometimes."

One of the contestants asked him how Watson "heard"

the information. "It reads," Ferrucci said. "When the clue hits your retina, it hits Watson's chips." Another contestant wondered about the algorithms Watson used to analyze the different answers. "Can you tell us how it generates confidence scores?"

"I could tell you," Ferrucci said, clearly warming to the competitive nature of the Challenge, "but I'd have to shoot you."

For these sparring rounds, IBM hired a young and telegenic host named Todd Crain. An actor originally from Rockford, Illinois, Crain had blond hair, a square jaw, and a quick wit, and had acted in comedy videos for TheOnion.com. At IBM's *Jeopardy* studio, he mastered a fluid and hipper take on Alex Trebek. Unlike Ferrucci's scientists, who usually referred to Watson as a thing, Crain always addressed Watson as a person. Watson was a character he could relate to, an information prodigy who committed the stupidest and most hilarious errors imaginable. Crain encouraged the machine, flattered it, and upbraided it. Sometimes he closed his eyes theatrically and moaned, "Oooooh, Watson!"

Crain had fallen into the *Jeopardy* gig months earlier through a chance encounter with David Shepler. Crain was working on a pilot documentary called EcoFreaks, telling the stories of people working at the fringes of the environmental movement. He said it involved spending one evening in New York with "freegans," Dumpster-divers devoted to reusing trash. On the next assignment, Crain and the crew drove north to the college town of New Paltz, New York. There Shepler—with the attention to detail he later demonstrated managing the *Jeopardy* project—had built a three-story house that would generate as much energy as it consumed, a so-called zero net-energy structure. While showing Crain the tri-ple-pane windows, geothermal exchange unit, and solar panel inverter, Shepler asked the young actor if he might be inter-ested in hosting a series of *Jeopardy* shows. "I said 'yes' before he even had a chance to finish the sentence," Crain said.

On occasion, Crain could irritate Ferrucci by making jokes at Watson's expense. The computer often opened itself to such jibes by mauling pronunciation, especially of foreign words. And it had the unfortunate habit of spelling out punc-tuation it didn't understand. One day, in the category Hair-y Situation, Watson said, "Let's play Hair-dash-Y Situation for two hundred." Crain imitated this bionic voice, getting a laugh from the small entourage of technicians and scientists. Ferrucci shook his head and muttered. Later, when Crain imi-tated a mangled name, Ferrucci channeled his irritation into feigned outrage: "He's making fun of him! It's like making fun of someone with a speech impediment!" (Once, Ferrucci said, he brought his wife and two daughters to a sparring session. When one of the girls heard Crain mimicking Watson, she said, "Daddy, why is that man being so *mean* to Watson?")

As the day's sparring session began, Crain gave the first two human contestants a chance to acquaint themselves with the buzzers. They tried several dozen old clues. Then he asked if they wanted Watson to join them. They nodded. "Okay, Burn, let him loose," Crain said. Burn Lewis, the member of Ferrucci's team who orchestrated the show from a tiny control room, pressed some buttons. The third competitor, an empty presence at the podium bearing the nameplate Watson, as-sumed its position. It might as well have been a ghost.

In the first game, it was clear the humans were dealing with a prodigious force that behaved differently from them. While humans almost always oriented themselves in a category by starting with the easier $200 clues, Watson began with the $1,000 clues at the bottom of the board and worked its way up. There was a logic to this. While humans heard all the answers, right and wrong, and learned from them, Watson was deaf to the proceedings. If it won the buzz, answered the clue, and got to pick another one, it could assume that it had been right. But that was its only feedback. Watson was sense-less to all of the signals its human competitors displayed — the smiles, the gasps, the confident tapping of fingers, the halt-ing speech and darting eyes spelling panic. More important, it lost out on game intelligence. If a human answered a clue incorrectly, Watson was liable to buzz on what was known as the rebound and deliver the very same incorrect answer. What was worse, Watson had a far harder time orienting it-self to the categories. How would it understand the Hair-y Situation category without hearing the other contestants' cor-rect answers? During these weeks, Ferrucci's team was talk-ing with the *Jeopardy* executives about giving Watson an elec-tronic feed with the text of the correct answer after each clue. But for the time being, the machine was learning nothing. So why not start each category with the priciest clues? The high bets might spook the humans. What's more, IBM's statistical analysis indicated that Watson was likely to land more Daily Doubles in those pricier realms.

Watson started off with Capital Cities, an apparently straightforward category that seemed to promise the ma-chine's favorite type of answer: factoids. It jumped straight to the $1,000 clue: "The Rideau Canal separates this North American capital into upper and lower regions." Todd Crain read the clue, and Lewis, in the control room, hit the button to turn on the light around the clue, opening it up for buzzes. Within milliseconds Watson had the clue all to itself.

"Watson?" Crain said.

"What is Ottawa?" Watson answered. With that, it raced through the entire category, with each correct answer reinforcing its confidence that it would know the others. Crain read each clue, the humans squeezed the button, and Watson beat them to it. It had no trouble with the South American city founded in 1535 by Pizarro ("What is Lima?") or the capital that fell to British troops in 1917 and to U.S. troops on April 9, 2003 ("What is Baghdad?"). These were factoids, each one wrapped in the most helpful data for Watson: hard facts unencumbered by humor, slang, or the cultural references that could tie a cognitive engine into knots. No, this category delivered a steady stream of dates, distances, specific names and numbers. For a *Jeopardy* computer, it was comfort food.

"Very good, Watson!" Crain said.

But after that promising start, Watson started to falter. Certain categories were simply hard for it to figure out. One was called I'll Take the Country from Here, Thanks. When Watson processed the $400 clue, "Nicolas Sarkozy from Jacques Chirac," it didn't know how to answer. In a few milliseconds it could establish that both names corresponded to presidents of France. But it did not understand the category well enough to build up confidence in an answer ("What is France?"). And it was not getting any orientation from the action of the game. Humans owned that category. Watson sat it out.

Then, in the category Collegiate Rhyme Time, Watson showed its stuff, but not enough to win. One asked for the smell of the document you receive upon graduating. Watson understood the clue perfectly and searched for synonyms for "document," then attempted to match them with words related to "smell." The best it could come up with was "What is bill feel?" ("What is diploma aroma?").

The real problems started when Watson found itself facing Greg Lindsay, a journalist and a two-time *Jeopardy* champion. Lindsay, thirty-two, had spent much of his time at the University of Illinois on the Quiz Bowl circuit, where he occasionally ran into Ken Jennings. In order to spar with Watson, Lindsay had to sign David Shepler's nondisclosure agreement. IBM wanted to keep Harry Friedman and his minions in the dark, as much as possible, about Watson's strengths and vulnerabilities. And Friedman didn't want the clues escaping onto the Internet before they aired on television. This meant that even if Lindsay defeated Watson, he wouldn't be able to brag about it to the Quiz Bowl community. For his crowd, this would be the equivalent of besting Kobe Bryant in a one-on-one game of hoops, then having to pretend it hadn't happened.

Even so, Lindsay came with a clear strategy to defeat Watson. He quickly saw that Watson mastered factoids but struggled with humor and irony, so he steered clear of Watson-friendly categories. He figured Watson would clean up on Name that Continent, picking out the right landmasses for Estado de Matto Grosso ("What is South America?") and the Filchner Ice Shelf ("What is Antarctica?"). The category Superheroes Names through Pictures looked much more friendly to humans. Sure enough, Watson was bewildered by clues such as "X marks the spot, man, when this guy opens his peeper" ("What is cyclops?"). Band Names also posed problems for Watson because the clues, like this one, were so murky: "The soul of a deceased person, thankful to someone for arranging his burial" ("What is the Grateful Dead?"). If the clue had included the lead guitarist Jerry Garcia or a famous song by the band, Watson could have identified it in an instant. But clues based on allusions, not facts, left it vulnerable.

More important, since the currency they were playing with was worthless, Lindsay decided to bet the maximum on each Daily Double. If he blew it, he lost nothing. And since he wasn't on national television, his reputation wouldn't suffer. As he put it, "There's no societal fear." Yet if he won his big bets, he'd be positioned to withstand Watson's inevitable charges through categories it understood. "I knew he would go on tears," Lindsay said. "I had to build up big leads when I had the chance." He aced his big bets and ended up thrashing Watson three times, once scoring an astronomical $59,999 of funny money. (The *Jeopardy* single-game record was $52,000 until Ken Jennings crushed it, winning $75,000 in his thirty-eighth game.)

Fortunately for Lindsay, he got Watson on what soon appeared to be a very bad day for the bionic star. The speech defect returned. When naming "one of the two monarchies that border China," the computer said, "What is Bhutand?" The game judge, Karen Ingraffea, consulted with David Shepler. From the observation room, Ferrucci could see them talking but could not hear a word. Shepler nodded grimly. Then he delivered the verdict to Todd Crain. Again Watson was docked, this time $1,000.

"This is silliness!" Ferrucci said.

His concern deepened as Watson started to strike out on questions that should have been easy. One Final Jeopardy clue, in the category 20th-Century People, looked like a cinch. It said: "The July 1, 1946, cover of *Time* magazine featured him with the caption, 'All matter is speed and flame'" ("Who is Albert Einstein?"). Watson displayed its top answers on its electronic panel. They were all ridiculous, and to the machine's credit, it had rock-bottom confidence in them. First was Time 100, a list of influential people that at one time included Einstein. But Watson should have known that the clue was ask-ing for a "him," not an "it." For more than two years, Ferruc-ci's language programmers had been preparing the machine to parse these clues. They had defined and mapped the twenty-five hundred things *Jeopardy* clues ask about. The most com-mon of these LATs were looking for a male person, a "he." De-termining that this clue was asking for a man's name should not have been so hard.

Watson's second choice, even more absurd, was David Ko-resh, the founder of the apocalyptic Branch Davidian cult near Waco, Texas. Koresh appeared on the May 3, 1993, cover of *Time,* days after burning down his compound and immo-lating everyone in it, including himself. No doubt the "flame" in the clue led Watson to Koresh. But Koresh was not born until thirteen years after Einstein appeared on the *Time* cover. Watson's other stabs were "stroke" and the painter Andrew Wyeth.

At this point, Ferrucci's frustration boiled over. He wasn't so bothered by the wild guesses, like David Koresh. The sys-tem had come up with a few answers that were somehow con-nected to the clue—a common magazine cover or flame. The confidence engine had done its job. After studying them, it had found little to go on and declared them worthless. "Wat-son's low-confidence answers are just garbage," Ferrucci had told the contestants earlier.

But why didn't Watson find the right answer? For a com-puter with access to millions of documents and lists, the July 1, 1946, cover profile in the nation's leading newsmagazine shouldn't be a deep mystery.

Ferrucci concluded that something was wrong with Wat-son and he wanted the team in the War Room at Hawthorne to get working on it right away. Yet even in one of the world's leading technology companies, it wasn't clear how to send the digital record of the computer's misadventures through the Internet. Ferrucci asked Eric Brown, then Eddie Epstein, and then Brown again: "How do I get the xml file to Hawthorne?" For Ferrucci, this failed game was brimming with vital feedback. It could point the Hawthorne team toward crucial fixes. The idea that his team could not respond immediately to whatever ailed Watson filled him with dread. Just imagine if Watson reprised this disastrous performance in its nationwide debut with Jennings and Rutter. "HOW DO I GET THIS FILE TO HAWTHORNE?" he shouted. No one had a quick answer. Ferrucci continued to thunder while, on the other side of the window, Todd Crain, Watson, and the other *Jeopardy* players blithely continued their game. (Watson, for one, was completely unfazed.) Finally Brown confirmed that he could plug a thumb drive into one of Watson's boxes, download the game data, and e-mail it to the team in Hawthorne. It promised to be a long night in the War Room, as the researchers diagnosed Watson's flops and struggled to restore its cognitive mojo.

Cloistered in a refrigerated room on the third floor of the Hawthorne labs stood another version of Watson. It turned out that the team needed two Watsons: the game player, engineered for speed, and this slower, steadier, and more forgiving system for development. The speedy Watson, its algorithms deployed across more than 2,000 processors, was a finicky beast and near impossible to tinker with. This slower Watson kept running while developers rewrote certain instructions, shifted out one algorithm for another, or refined its betting strategy. It took forty minutes to run a batch of questions, but it could handle two hundred at a time. Unlike the fast machine, it created meticulous records, and it permitted researchers to experiment, section by section, with its answering process. Because the team could fiddle with the slower machine, it was always up-to-date, usually a month or two ahead of its speedy sibling. After the debacle against Lindsay, IBM could only hope that the slower, smarter Watson wouldn't have been so confused.

Within twenty-four hours, Ferrucci's team had run all of that day's games on the slow machine. The news was encouraging. It performed 10 percent better on the clues. The biggest difference, according to Eric Brown, was that some of the clues were topical, and speedy Watson's most recent data came from 2008. "We got creamed on a couple categories that required much more current information," he said.

Other recent adjustments in the slow Watson helped it deal with chronology. Keeping track of facts as they change over time is a chronic problem for AI systems, and Watson was no exception. In the recent sparring session, it had confused a mid-nineteenth-century novel for a late-twentieth-century pop duo. Yet when Ferrucci analyzed the slower Watson's performance on the problematic Oliver Twist clue, he was relieved to see that a recent tweak had helped the machine match the clue to the right century. This fix in "temporal reasoning" pushed the Pet Shop Boys answer way down its list, from first to number 79. Watson's latest top answer—"What is magician?"—was still wrong but not as laughable. "It still knows nothing about Oliver Twist," Ferrucci wrote in a late-night e-mail.

While Ferrucci and a handful of team members attended every sparring match in the winter of 2010, Jennifer Chu-Carroll generally stayed away. For her, their value was in the data they produced, not the spectacle, and much less the laughs. As she saw it, the team had a long list of improvements to make before autumn. By that point, the immense collection of software running Watson would be locked down—frozen. After that, the only tinkering would be in a few peripheral applications, like game strategy. But the central operations of the computer, like those of other mission-critical systems, would go through little but testing during the months leading up to the *Jeopardy* showdown. Engineers didn't dare tinker with Space Shuttle software once the vessel was headed toward the launch pad. Watson would get similar treatment.

With each sparring session, however, the list of fixes was getting longer. For each fix, the team had to weigh the time it would take against the possible gain in performance. "It's triage," Chu-Carroll said. During one sparring session, for example, Watson mispronounced *wienerschnitzel*, neglecting to say the "W" as a "V." Was it worth the trouble to fine-tune its German phonetics? Not unless someone could do it in a hurry.

In one Final Jeopardy, Watson inched closer to the fix-it threshold. Asked to identify the sole character in the American Film Institute's list of the fifty greatest heroes who was not portrayed by a human, the computer came back with "Who is Buffy the Vampire Slayer?" The audience laughed, and Todd Crain slapped his forehead, saying, "Oh, Watson, for the love of God!"

Still, solving that clue would have been a formidable challenge. Once Watson found the list of heroes, it would have had to carry out fifty separate searches to ascertain that each of the characters, from Atticus Finch to James Bond, Indiana Jones, and *Casablanca*'s Rick Blaine, was human. (It wouldn't necessarily be that easy, since most documents and databases don't note a protagonist's species.) During that search, presumably, it would see that thirty-ninth on the list was a collie, a breed of dog (and therefore not human), and would then display "Who is Lassie?" on its electronic screen. Would the lessons gained in learning how to spot the dog in a long list of humans pay off elsewhere? Probably not.

That raised another question for the harried team. If Watson had abysmally low confidence in a Final Jeopardy response, as was the case with the Pet Shop Boys and Buffy the Vampire Slayer, would it be better to say nothing? If it was in the company's interest to avoid looking stupid, suppressing wild guesses might be a good move. This dilemma did not arise with the regular *Jeopardy* clues. There, if Watson lacked confidence in an answer, it simply refrained from buzzing. But in Daily Doubles and Final Jeopardy, contestants had to bet before seeing the clue. Humans guessed when they didn't know the answer. This is what Watson was doing, too. But its chronic shortage of common sense made its guesses infinitely dumber. In the coming weeks, the IBM team would calculate the odds of a lucky guess for each of Watson's confidence levels. While *Jeopardy* executives, eager for entertainment and high ratings, would no doubt favor the occasional outrageous guess, IBM had other priorities. "At low levels of confidence, I think we'll just have it say it doesn't know," said Chu-Carroll. "Sometimes that sounds smarter."

Mathematics was one category where the IBM machine could not afford to look dumb. The company, after all, was built on math. However, the *Jeopardy* training data didn't include enough examples to educate Watson in this area. Of the more than seventy-five thousand clues Eric Brown and his team studied, only fourteen involved operations with fractions. A game strategist wouldn't dwell on them. But for IBM, there was more at risk than winning or losing a game. To prepare Watson for math, the team might have to put aside the statistical approach and train the machine in the rules and lingo of arithmetic.

As they worked to lift Watson's performance, the *Jeopardy* team focused on entire categories that the machine misunderstood. They called them train wrecks. It was a new genre, conceived after Watson's debacle against Lindsay. The most insidious train wrecks, Gondek said one afternoon, were those in which Watson was fooled into "trusting" its expertise—generating high confidence scores—in categories where it in fact had no clue. This double ignorance could lead it to lay costly bets, embarrassing the team and losing the match.

Lots of the train wreck categories raised questions about the roots of Watson's misunderstandings. One category, for example, that appeared to confuse it was Books in Español. Watson didn't come close to identifying Ernest Hemingway's *Adios a las Armas,* Harper Lee's *Matar un Ruiseñor,* or Stephen King's *La Milla Verde.* It already held rudimentary foreign words and phrases in its tool kit. But would it benefit from greater detail? As it turned out, Watson's primitive Spanish wasn't the problem. The issue was simpler than that. From the name of the category and the bare-bones phrasing of the clues—Stephanie Meyer: *Luna Nueva*—the computer did not know what to look for. And unlike human contestants, it was deaf to the correct answers. If IBM and *Jeopardy* ironed out an arrangement to provide Watson with the answers after each clue, it might orient itself in puzzling categories. That way, it could move on to the real challenge of the clue, recognizing titles like *To Kill a Mockingbird* and *A Farewell to Arms* in Spanish.

· · ·

As the season of sparring sessions progressed, people in the observation room paid less attention to the matches as they were being played. They talked more and looked up at the big monitor when they heard laughter or when Watson found itself in a tight match. The patterns of the machine were becoming familiar. For them, much of the excitement came a day later, when they began to analyze the data and saw how the smarter version of Watson handled the troublesome clues. Ferrucci occasionally used the time during the matches to explain Watson's workings to visitors, or to give interviews. One March morning, he could be heard across the room talking to a documentary producer. Asked if he would be traveling to California for the televised final match, Ferrucci deadpanned: "I'll be sedated."

David Gondek, sitting across from Ferrucci, his fingers on his laptop keyboard, said that pressure in the War Room was mounting. He had largely abandoned his commute from Brooklyn and now spent nights in a small apartment he'd rented nearby. It was only ten minutes by bike to the War Room or a half hour to pedal to the Yorktown labs, where the sparring sessions took place.

From the very beginning, Gondek said, the *Jeopardy* challenge differed from a typical software project. Usually, software developers are given a list of functions and applications to build. And when they finish them, test, tweak, and debug them, they're done. Building Watson, however, never ended, he said. There was always something it failed to understand. The work, he said, "is infinite."

In graduate school, Gondek had focused on data mining. His thesis, on nonredundant clustering, involved programming machines to organize clusters of data around connections that the users might not have considered. By answering some preliminary questions, for example, an intelligence officer might inform the system that he's all too familiar with Osama bin Laden's connections to terrorism. So the system, when sorting through a batch of intelligence documents, would find other threads and connections, perhaps leading to fresh insights about the Al Qaeda leader. Machines, much like humans, follow conventional patterns of analysis. Gondek had been thinking about this since listening to a recent talk by a cognitive psychologist. It raised this question: If a machine like Watson fell into the same mental traps as humans, was it a sign of intelligence or just a cluelessness that it happened to share with us? He provided an example.

"What color is snow?" he asked.

"White," I said.

"A wedding dress?"

"White."

"Puffy clouds?"

"White."

"What do cows drink?"

"Milk," I said, falling obediently into the trap he'd set.

Cows, of course, drink water once they're weaned. Because humans naturally seek patterns and associations, most of us get into a "white" frame of mind. Psychologists call this the associative network theory. One node in our mind represents "cow," said Penn State's Richard Carlson. "It's related to others, for milk and steak and mooing, and so on." The mention of "cow," he said, activates the entire network, priming it. "That way, you're going to be quicker to respond."

Gondek's point was that Watson, unlike most question-answering programs, would fall for the same trick. It focused on patterns and correlations and had a statistical version of an associative network. It was susceptible to being primed for "white." It was like a human in that narrow way.

University researchers in psychology and computational neuroscience are building computer models to probe these similarities. At Carnegie Mellon, a team under John Anderson, a psychology professor, has come up with a cognitive architecture called ACT-R that simulates human thought processes. Like Watson, it's a massively parallel system fueled by statistical analysis.

Yet the IBM team resolutely avoided comparisons between Watson's design and that of a brain. Any claims of higher intelligence on the part of their machine, they knew, would provoke a storm of criticism from psychologists and the AI community alike. It was true that on occasion Watson and the human brain appeared to follow similar patterns. But that, said Gondek, was only because they were programmed, each in its own way, to handle the same job.

A few months later, Greg Lindsay was eating sushi in a small Japanese restaurant near his apartment in Brooklyn Heights. He wore wire-rimmed glasses, and his thinning hair was cut so short that it stood straight up. He had to eat quickly. A book editor was waiting for the manuscript fixes on his book, *Aerotropolis*. It was about the rise of cities built around airports, and it fit his insatiable hunger for facts. He said he had had to delve deeply into transportation, energy, global manufacturing, and economics. Little surprise, then, that the book was nearly five hundred pages long.

Lindsay said he had assumed that Watson would maul him in the sparring rounds, beating him to the buzzer every time. This hadn't happened. By anticipating the end of Todd Crain's delivery of the clue, he had managed to outbuzz Watson a number of times. He also thought that the extra time to answer Final Jeopardy would give Watson plenty of opportunity to find the right answer. This clearly was not the case. In fact, this extra time raised questions among Ferrucci's team. To date, Watson was answering every question the same way, as if it had the usual three to five seconds, even when it had five or six times as long. That meant that it was forgoing precious time that it could be spending hunting and evaluating potential answers. Would that extra time help? Just a few days before, Gondek had said that he wasn't sure. With more time, he said, "Watson might bring more wrong answers to the surface and undermine its confidence in the right one." In other words, the computer ran the risk of overthinking. In the coming months, Gondek and his colleagues thought they might test a couple of other approaches, but they were starting to run out of time.

For his strategy against Watson, Lindsay said, he took a page out of the William Goldman novel *The Princess Bride*. The hero in the story is facing a fight with a much better swordsman, so he contrives to move the fight to a stony surface, where the rival might slip. In the same way, Lindsay steered Watson to an equally unstable arena: "areas of semantic complexity." He predicted that humans playing Watson in the television showdown would follow the same strategy.

But there was one big difference. With a million dollars at stake, the humans would not only be battling Watson, they'd also be competing against each other. This could change the dynamics dramatically. In the sparring sessions, the humans (playing with funny money) focused exclusively on the machine. "I didn't care about the others; I just wanted to beat Watson," Lindsay said. But as the two humans in the upcoming match probed each other's weaknesses and raced to buzz peremptorily, they could open the door for the third contestant, who would be oblivious to the drama and would go about its business, no doubt, with the unflappable dispatch of a machine.

7

AI

ON A MIDSUMMER afternoon in 2010, a cognitive scientist at MIT named Joshua Tenenbaum took a few minutes to explain why the human brain was superior to a question-answering machine like Watson. He used the most convenient specimen of human cognition at hand, his own mind, to make his case. Tenenbaum, a youthful professor with sandy hair falling across his forehead and an easy smile, has an office in MIT's imposing headquarters for research on brains, memory, and cognitive science. His window looks across the street at the cascading metallic curves of MIT's Stata Center, designed by the architect Frank Gehry.

Tenenbaum is focusing his research on the computational basis of human learning and trying to replicate it with machines. His goal is to come up with computers whose intelligence reaches far beyond answering questions or finding correlations in masses of data. One day, he hopes, the systems he's working on will come up with concepts and theories, the way humans do, sometimes basing them on just a handful of observations. They would make what he called inductive leaps, behaving more like Charles Darwin than, say, Google's search engine or Watson. Darwin's data—his studies of worms, pigeons, and a host of other plants and animals—was tiny by today's standards; they would occupy no more than a few megabytes on a hard drive. Yet he came up with a theory that explained the evolution of life on earth. Could a computer do that?

Tenenbaum was working toward that distant vision, but for the moment his objective was more modest. He thought Watson acted smarter than it was, and he wanted to demonstrate why. He had recently read in a magazine about Watson's mastery of *Jeopardy*'s Before and After clues, the ones that linked two concepts or people with a shared word in the middle. When asked about a candy bar that was a Supreme Court justice, Watson had quickly come up with "Who is Baby Ruth Ginsburg."

Now Tenenbaum was creating a Before and After clue of his own. "How about this one?" he said. "A president who wrote a founding document and later led a rebellion against it." The answer, a combination of the third president of the United States and the only president of the Confederacy: Thomas Jefferson Davis.

Tenenbaum's point was that it took a team of gifted engineers to teach Watson how to handle these questions by devising clever algorithms. But humans, after seeing a single example of a Before and After clue, could build on it, not only figuring out how to respond to such questions but inventing new ones. "I know who Ruth Ginsburg is and I know what Baby Ruth is and I see how they overlap, and from that one example I can extract that template," he said. "I don't have to be programmed with that question." We humans, he explained, create our own algorithms on the fly.

As in many fields of science, researchers in Artificial Intel-

ligence have long fallen into two groups, pragmatists and visionaries. And most of the visionaries, including Tenenbaum,
argue that machines like Watson merely simulate intelligence
by racing through billions of correlations. Watson and its kin
don't really "know" or "understand" anything. Watson can ace
Jeopardy clues on Shakespeare, but only because the ones and
zeros that spell out "Shakespeare" pop up on lists and documents near other strings of ones and zeros representing play-
wrights, England, Hamlet, Elizabethan, and so on. It lacks
anything resembling awareness. Most reject the suggestion
that the clusters of data nestled among its transistors mirror
the memories encoded chemically in the human brain or that
Watson's search for *Jeopardy* answers, and its statistical methods of balancing one candidate answer with another, mimic
what goes on in Ken Jennings's head.

The parallels, Tenenbaum said, are deceiving. Watson, for
example, appears to learn. But its learning comes from adjusting its judgments to feedback, moving toward the combinations that produce correct answers and away from errors.
These "error-driven learning algorithms," he said, are derived
from experiments in behavioral psychology. "The animals do
something, and they're rewarded or they're punished," he said.
That kind of learning may be crucial to survival, leading humans and many animals alike to recoil from flames, coiled
snakes, and bitter, potentially poisonous, berries. But this describes a primitive level of brain function. What's more, Watson's learning laboratory was limited, extending only to its 75
gigabytes of data and the instructions of its algorithms. Outside that universe, Tenenbaum stressed, Watson knew nothing. And it formed no theories.

Ferrucci didn't disagree. Watson had its limitations. One

time, when Ferrucci learned that another scientist had dispar-
aged Watson as an "idiot savant," he said, "Idiot savant? I'll
take it!" While he objected to that term, which he viewed as
demeaning, Ferrucci said he only wished that Watson could
approach the question-answering mastery of humans like
Kim Peek, the model for the so-called megasavant played by
Dustin Hoffman in the movie *Rainman*. Peek, who died in
2009, was a walking encyclopedia. He had read voluminously
and seemed to recall every detail with precision. Yet he had
grave physical and developmental shortcomings. His brain
was missing the corpus callosum, the bundle of nerves connecting the two hemispheres. He had little meaningful inter-
action with people—with the exception of his father—and
he did not appear to draw sophisticated conclusions from his
facts, much less come up with theories. He was a stunted genius. But unlike Watson, he was entirely fluent in language.
As far as Ferrucci was concerned, a Q-A machine with the
language proficiency of a human was a dream. It would have
boundless market potential. He would leave it to other kinds
of machines to come up with theories.

The question was whether computers like Watson, products of this pragmatic, problem-solving (and profit-seeking)
side of the AI world, were on a path toward higher intelligence. Within a decade, computers would likely run five hundred times as fast and would race through databases a thousand times as large. Within fifteen years, studies predicted
that a single supercomputer would be able to carry out 10^{20}
calculations per second. This was enough computing power
to count every grain of sand on earth in a single second (assuming it didn't have more interesting work to do). At the
same time, the algorithms running such machines, each one

resulting from decades of rigorous Darwinian sifting, would be smarter and more precise. Would these supercharged descendants of Watson still be in the business of simulating intelligence? Or could they make the leap to a human level, then advance beyond?

The AI community was full of doubters. And their concerns about the limitations of statistical crunchers like Watson stirred plenty of debate within the scientific community. Going back decades, the sparkling vision of AI was to develop machines that could think, know, and learn. Watson, many argued, landed its star spot on national television without accomplishing any of those goals. A human answering a *Jeopardy* question draws on "layers and layers of knowledge," said MIT's Sajit Rao. "There's so much knowledge around every single word." Watson couldn't compare. "If you ask Watson what time it is," wrote one computer scientist in an e-mail, "it won't have an answer."

If Watson hadn't been so big, few would have cared. But the size and scope of the project, and the razzmatazz surrounding it, fueled resentment. Big Blue was a leading force in AI, and its focus on *Jeopardy* funneled more research dollars toward its statistical approach. What's more, Watson was sure to hog the press. The publicity leading up to the man-machine *Jeopardy* showdown would likely shine a brighter spotlight on AI than anything since the 1997 chess matches between Garry Kasparov and Deep Blue. Yet the world would see, and perhaps fall in love with, a machine that only *simulated* intelligence. In many aspects, it was dumb. And despite its mastery of statistics, it knew nothing. Worse, if Watson—despite these drawbacks—proved to be an effective and versatile knowledge machine, it might spell the end of competing technologies, turning years of research—entire careers—into dead ends.

The final irony: At least a few scientists kept their criticism of Watson private for fear of alienating Big Blue, a potential sponsor of their research.

In sum, from a skeptic's view, the machine was too dumb, too ignorant, too famous, and too rich. (In that sense, IBM's computer resembled lots of other television stars. And, interestingly enough, the resentment within the field mirrored the combination of envy and contempt that serious actors feel for the celebrities on reality TV.)

These shortcomings aside, Watson had one quality that few could ignore. In the broad realm of *Jeopardy*, it worked. It made sense of most of the clues, even those in complex English, and it came up with answers within a few seconds. The question was whether other lines of research in AI would surpass it—or perhaps one day endow a machine with the human smarts or expertise that it lacked.

Dividing the pragmatists like Ferrucci and the idealists within AI was the human brain. For many, including Tenenbaum, the path toward true machine intelligence had less to do with the power of the computer than the nature of its instructions and architecture. Only the brain, they believed, held the keys to higher levels of thinking—to concepts, ideas, and theories. But they were tangled up in the most complex circuitry known in the universe.

Tenenbaum compared the effort required to build theorizing and idea-spouting machines with the American push, a half century earlier, to send a manned voyage to the moon. The moon shot, he said, was far easier. When President Kennedy issued his call for a lunar mission in May 1961, most of the basic scientific research had already been accomplished. Indeed, the march toward space travel had begun early in the seventeenth century, when Galileo started to write down the mathematical equations describing how certain objects moved. This advanced through the Scientific and Industrial Revolutions, from the physics of Newton to the harnessing of electricity, the development of chemical bonds and powerful fuels, the creation of metal alloys, and, finally, advances in rocket technology. By the 1960s, the basic science behind sending a spaceship to the moon was largely complete. Much of the technology existed. It was up to the engineers to assemble the pieces, build them to the proper scale, and send the finished spacecraft skyward.

"If you want to compare [AI] to the space program, we're at Galileo," Tenenbaum said. "We're not yet at Newton." He is convinced that while ongoing research into the brain is shining a light on intelligence, the larger goal—to reverse engineer human thought—will require immense effort and time. An enormous amount of science awaits before the engineering phase can begin. "The problem is exponentially harder [than manned space flight]," he said. "I wouldn't be surprised if it took a couple hundred years."

No wonder, you might say, that IBM opted for a more rapid approach. Yet even as Tenenbaum and others engage in quasi-theological debates about the future of intelligence, many are already snatching ideas from the brain to build what they can in the here-and-now. Tenenbaum's own lab, using statistical formulas inspired by brain functions, is training computers to sort through scarce data and make predictions about everything from the location of oil deposits to suicide bombing attacks. For this, he hopes to infuse the machines with a thread or two of logic inspired by observations of the brain, helping them to connect dots the way people do. At the same time, legions of theorists, focused on the exponential advances in computer technology, are predicting that truly smart machines, also inspired by the brain, will be arriving far ahead of Tenenbaum's timetable. They postulate that within decades, computers more intelligent than humans will dramatically alter the course of human evolution.

Meanwhile, other scientists in the field pursue a different type of question-answering system—a machine that actually knows things. For two generations, an entire community in AI has tried to teach computers about the world, describing the links between oxygen and hydrogen, Indiana and Ohio, tables and chairs. The goal is to build knowledge engines, machines very much like Watson but capable of much deeper reasoning. They have to know things and understand certain relationships to come up with insights. Could the emergence of a data-crunching wonder like Watson short-circuit their research? Or could their work help Watson grow from a dilettante into a scholar?

In the first years of the twenty-first century, Paul Allen, the cofounder of Microsoft, was pondering Aristotle. For several decades in the fourth century BC, that single Greek philosopher was believed to hold most of the world's scientific knowledge in his head. Aristotle was like the Internet and Google combined. He stored the knowledge and located it. In a sense, he outperformed the Internet because he combined his factual knowledge with a mastery of language and context. He could answer questions fluently, and he was reputedly a wonderful teacher.

This isn't to say that as an information system, Aristotle had no shortcomings. First, the universe of scientific knowledge in his day was tiny. (Otherwise it wouldn't have fit into one head, no matter how brilliant.) What's more, the bandwidth in and out of his prodigious mind was severely limited. Only a small group of philosophers and students (including the future Alexander the Great) enjoyed access to it, and then only during certain hours of the day, when the philosopher turned his attention to them. He did have to study, after all. Maintaining omniscience—or even a semblance of it—required hard work.

For perhaps the first time since the philosopher's death, as Allen saw it, a single information system—the Internet —could host the universe of scientific knowledge, or at least a big chunk of it. But how could people gain access to this treasure, learn from it, winnow the truth from fiction and innuendo? How could computers teach us? The solution, it seemed to him, was to create a question-answering system for science, a digital Aristotle.

For years, Allen had been plowing millions into research on computing and the human brain. In 2003, he directed his technology incubator, Vulcan Inc., of Seattle, to sponsor long-range research to develop a digital Aristotle. The Vulcan team called it Project Halo. This scientific expert, they hoped, would fill a number of roles, from education to research. It would answer questions for students, maybe even develop a new type of interactive textbook. And it would serve as an extravagantly well-read research assistant in laboratories.

For Halo to succeed in these roles, it needed to do more than simply find things. It had to weave concepts together. This meant understanding, for example, that when water reaches 100 degrees centigrade it turns into steam and behaves very differently. Plenty of computers could impart that information. But how many could incorporate such knowledge into their analysis and reason from it? The idea of Halo was to build a system that, at least by a liberal definition of the word, could think.

The pilot project was to build a computer that could pass the college Advanced Placement tests in chemistry. Chemistry, said Noah Friedland, who ran the project for Vulcan, seemed like the ideal subject for a computer. It was a hard science "without a lot of psychological interpretations." Facts were facts, or at least closer to them in chemistry than in squishier domains, like economics. And unlike biology, in which tissue scans and genomic research were unveiling new discoveries every month or two, chemistry was fairly settled. Halo would also sidestep the complications that came with natural language. Vulcan let the three competing companies, two American and one German, translate the questions from English into a logic language that their systems could understand. At some point in the future, they hoped, this digital Aristotle would banter back and forth in human languages. But in the four-month pilot, it just had to master the knowledge and logic of high school chemistry.

The three systems passed the test, albeit with middling scores. But if you looked at the process, you'd hardly know that machines were involved. Teaching chemistry to these systems required a massive use of human brainpower. Teams of humans—knowledge engineers—had to break down the fundamentals of chemistry into components that the computers could handle. Since the computer couldn't develop concepts on its own, it had to learn them as exhaustive lists and laws. "We looked at the cost, and we said, 'Gee, it costs $10,000 per textbook page to formulate this knowledge,'" Friedland said.

It seemed ludicrous. Instead of enlisting machines to help sort through the cascades of new scientific information, the machines were enlisting humans to encode the tiniest fraction of it—and at a frightful cost. The Vulcan team went on to explore ways in which thousands, or even millions, of humans could teach these machines more efficiently. In their vision, entire communities of experts would educate these digital Aristotles, much the way online communities were contributing their knowledge to create Wikipedia. Work has continued through the decade, but the two principles behind the Halo thinking haven't changed: First, smart machines require smart teachers, and only humans are up to the job. Second, to provide valuable answers, these computers have to be fed factual knowledge, laws, formulas, and equations.

Not everyone agrees. Another faction, closely associated with search engines, is approaching machine intelligence from a different angle. They often remove human experts from the training process altogether and let computers, guided by algorithms, study largely on their own. These are the statisticians. They're closer to Watson's camp. For decades, they've been at odds with their rule-writing colleagues. But their approach registered a dramatic breakthrough in 2005, when the U.S. National Institute for Standards and Technologies held one of its periodic competitions on machine translation. The government was ravenous for this translation technology. If machines could automatically monitor and translate Internet traffic, analysts might get a jump on trends in trade and technology and, even more important, terrorism. The competition that year focused on machine translation from Chinese and Arabic into English. And it drew the usual players, including a joint team from IBM and Carnegie Mellon and a handful of competitors from Europe. Many of these teams, with their blend of experts in linguistics, cognitive psychology, and computer science, had decades of experience working on translations.

One new player showed up: Google. The search giant had been hiring experts in machine translation, but its team differed from the others in one aspect: No one was expert in Arabic or Chinese. Forget the nuances of language. They would do it with math. Instead of translating based on semantic and grammatical structure, the interplay of the verbs and objects and prepositional phrases, their computers were focusing purely on statistical relationships. The Google team had fed millions of translated documents, many of them from the United Nations, into their computers and supplemented them with a multitude of natural-language text culled from the Web. This training set dwarfed their competitors'. Without knowing what the words meant, their computers had learned to associate certain strings of words in Arabic and Chinese with their English equivalents. Since they had so very many examples to learn from, these statistical models caught nuances that had long confounded machines. Using statistics, Google's computers won hands down. "Just like that, they bypassed thirty years of work on machine translation," said Ed Lazowska, the chairman of the computer science department at the University of Washington.

The statisticians trounced the experts. But the statistically trained machines they built, whether they were translating from Chinese or analyzing the ads that a Web surfer clicked, didn't know anything. In that sense, they were like their question-answering cousins, the forerunners of the yet-to-be-conceived *Jeopardy* machine. They had no response to different types of questions, ones they weren't programmed to answer. They were incapable of reasoning, much less coming up with ideas.

Machines were seemingly boxed in. When people taught them about the world, as in the Halo project, the process was too slow and expensive and the machines ended up "overfit-ted"—locked into single interpretations of facts and relation-ships. Yet when machines learned for themselves, they turned everything into statistics and remained, in their essence, ig-norant.

How could computers get smarter about the world? Tom Mitchell, a computer science professor at Carnegie Mellon, had an idea. He would develop a system that, just like mil-lions of other students, would learn by reading. As it read, it would map all the knowledge it could make sense of. It would learn that Buenos Aires appeared to be a city, and a capital too, and for that matter also a province, that it fit inside Ar-gentina, which was a country, a South American country. The computer would perform the same analysis for billions of other entities. It would read twenty-four hours a day, seven days a week. It would be a perpetual reading machine, and by extracting information, it would slowly cobble together a net-work of knowledge: every president, continent, baseball team, volcano, endangered species, crime. Its curriculum was the World Wide Web.

Mitchell's goal was not to build a smart computer but to construct a body of knowledge—a corpus—that smart com-puters everywhere could turn to as a reference. This com-puter, he hoped, would be doing on a global scale what the human experts in chemistry had done, at considerable cost, for the Halo system. Like Watson, Mitchell's Read-the-Web computer, later called NELL, would feature a broad range of analytical tools, each one making sense of the readings from its own perspective. Some would compare word groups, oth-ers would parse the grammar. "Learning method A might de-cide, with 80 percent probability, that Pittsburgh is a city," Mitchell said. "Method C believes that Luke Ravenstahl is the

mayor of Pittsburgh." As the system processed these two beliefs, it would find them consistent and mutually reinforcing.
If the entity called Pittsburgh had a mayor, there was a good
chance it was a city. Confidence in that belief would rise. The
computer would learn.

Mitchell's team turned on NELL in January 2010. It
worked on a subsection of the Web, a cross section of two
hundred million Web pages that had been culled and curated
by Mitchell's colleague Jamie Callan. (Operating with a fixed
training set made it easier in the early days to diagnose troubles and carry out experiments.) Within six months, the machine had developed some four hundred thousand beliefs—a
minute fraction of what it would need for a global knowledge
base. But Mitchell saw NELL and other fact-hunting systems
growing quickly. "Within ten years," he predicted, "we'll have
computer programs that can read and extract 80 percent of
the content of the Web, which itself will be much bigger and
richer." This, he said, would produce "a huge knowledge base
that AI can work from."

Much like Watson, however, this knowledge base would
brim with beliefs, not facts. After all, statistical systems
merely develop confidence in facts as a calculation of probability. They believe, to one degree or another, but are certain of nothing. Humans, by contrast, must often work from
knowledge. Halo's Friedland (who left Vulcan to set up his
own shop in 2005) argues that AI systems informed by machine learning will end up as dilettantes, like Watson (at least
in its *Jeopardy* incarnation). A computer, he said, can ill afford
to draw conclusions about a jet engine turbine based on "beliefs" about bypass ratios or the metallurgy of titanium alloys.
It has to know those things.

So when it came to teaching knowledge machines at the

end of the first decade of the twenty-first century, it was a question of picking your poison. Computers that relied on human teachers were slow to learn and frightfully expensive to teach. Those that learned automatically unearthed possible answers with breathtaking speed. But their knowledge was superficial, and they were unable to reason from it. The goal of AI—to marry the speed and range of machines with the depth and subtlety of the human brain—was still awaiting a breakthrough. Some believed it was at hand.

In 1859, the British writer Samuel Butler sailed from England, the most industrial country on earth, to the wilds of New Zealand. There, for a few years, he raised sheep. He was as far away as could be, on the antipodes, but he had the latest books shipped to him. One package included the new work by Charles Darwin, *On the Origin of Species.* Reading it led Butler to contemplate humanity in an evolutionary context. Presumably, humans had developed through millions of years, and their rhythms, from the perspective of his New Zealand farm, appeared almost timeless. Like sheep, people were born, grew up, worked, procreated, died, and didn't change much. If the species evolved from one century to the next, it was imperceptible. But across the seas, in London, the face of the earth was changing. High-pressure steam engines, which didn't exist when his parents were born, were powering trains across the countryside. Information was speeding across Europe through telegraph wires. And this was just the beginning. "In these last few ages," he wrote, referring to machines, "an entirely new kingdom has sprung up, of which we as yet have only seen what will one day be considered the antediluvian prototypes of the race." The next step of human evolution, he wrote in an 1863 letter to the editor of a local newspaper, would be led by the progeny of steam engines, electric turbines, and telegraphs. Human beings would eventually cede planetary leadership to machines. (Not to fear, he predicted: The machines would care for us, much the way humans tended to lesser beings.)

> What sort of creature [is] man's next successor in the supremacy of the earth likely to be? We have often heard this debated; but it appears to us that we are ourselves creating our own successors; we are daily adding to the beauty and delicacy of their physical organisation; we are daily giving them greater power and supplying by all sorts of ingenious contrivances that self-regulating, self-acting power which will be to them what intellect has been to the human race. In the course of ages we shall find ourselves the inferior race.

Butler's vision, and others like it, nourished science fiction for more than a century. But in the waning years of the twentieth century, as the Internet grew to resemble a global intelligence and computers continued to gain in power, legions of technogeeks and philosophers started predicting that the age of machines was almost upon us. They called it the Singularity, a hypothetical time in which progress in technology would feed upon itself feverishly, leading to transformational change.

In August 2010, hundreds of computer scientists, cognitive psychologists, futurists, and curious technophiles descended on San Francisco's Hyatt hotel, on the Embarcadero, for the two-day Singularity Summit. For most of these people, programming machines to catalogue knowledge and answer questions, whether manually or by machine, was a bit pedestrian. They weren't looking for advances in technology that already existed. Instead, they were focused on a bolder challenge, the development of deep and broad machine intelligence known as Artificial General Intelligence. This, they believed, would lead to the next step of human evolution.

The heart of the Singularity argument, as explained by the technologists Vernor Vinge and Ray Kurzweil, the leading evangelists of the concept, lies in the power of exponential growth. As Samuel Butler noted, machines evolve far faster than humans. But information technology, which Butler only glimpsed, races ahead at an even faster rate. Digital tools double in power or capacity every year or two, whether they are storing data, transmitting it, or performing calculations. A single transistor cost $1 in 1968; by 2010 that same buck could buy a billion of them. This process, extended into the future, signaled that sometime in the third decade of this century, computers would rival or surpass the power and complexity of the human brain. At that point, conceivably, machines would organize our affairs, come up with groundbreaking ideas, and establish themselves as the cognitive leaders of the planet.

Many believed these machines were yet to be invented. They would come along in a decade or two, powered by new generations of spectacularly powerful semiconductors, perhaps fashioned from exotic nanomaterials, the heirs to silicon. And they would feature programs to organize knowledge and generate language and ideas. Maybe the hardware would replicate the structure of the human brain. Maybe the software would simulate its patterns. Who knew? Whatever its configuration, perhaps a few of Watson's algorithms or an insight from Josh Tenenbaum's research would find their way into this machinery.

But others argued that the Singularity was already well under way. In this view, computers across the Internet were al-

ready busy recording our movements and shopping prefer-
ences, suggesting music and diets, and replacing traditional
brain functions such as information recall and memory storage. Gregory Stock, a biophysicist, echoed Butler as he placed
technology in an evolutionary context. "Lines are blurring
between the living and the not-living, between the born and
the made," he said. The way he described it, life leapt every
billion years or so into greater levels of complexity. It started
with simple algaelike cells, advanced to complex cells and
later to multicellular organisms, and then to an explosion of
life during the Cambrian period, some five hundred fifty million years ago. This engendered new materials within earth's
life forms, including bone. Stock argued that humans, using
information technology, were continuing this process, creating a "planetary superorganism"—a joint venture between
our cerebral cortex and silicon. He said that this global intelligence was already transforming and subjugating us, much the
way our ancestors tamed the gray wolf to create dogs. He predicted that this next step of evolution would lead to the de-
mise of "free-range humans," and that those free of the support and control of the planetary superorganism would retreat
to back eddies. "I hate to see them disappear," he said.

The crowd at the Singularity Summit was by no means
united in these visions. A biologist from Cambridge University, Dennis Bray, described the daunting complexity of a single cell and cautioned that the work of modeling the circuitry
and behavior of even the most basic units of life remained
formidable. "The number of distinguishable proteins that a
human makes is essentially uncountable," he said. So what
chance was there to model the human brain, with its hundred
billion neurons and quadrillions of connections?

In the near term, it was academic. No one was close to replicating the brain in form or function. Still, the scientists at the conference were busy studying it, hoping to glean from its workings single applications that could be taught to computers. The brain, they held, would deliver its treasures bit by bit. Tenenbaum was of this school.

And so was Demis Hassabis. A diminutive thirty-four-year-old British neuroscientist, Hassabis told the crowd that technology wasn't the only thing growing exponentially. Research papers on the brain were also doubling every year. Some fifty thousand academic papers on neuroscience had been published in 2008 alone. "If you looked at neuroscience in 2005, or before that, you're way out of date now," he said. But which areas of brain research would lead to the development of Artificial General Intelligence?

Hassabis had followed an unusual path toward AI research. At thirteen, he was the highest ranked chess player of his age on earth. But computers were already making inroads in chess. So why dedicate his brain, which he had every reason to believe was exceptional, to a field that machines would soon conquer? (From the perspective of futurists, chess was an early sighting of the Singularity.) Even as he played chess, Hassabis said later, he was interested in what was going on in his head—and how to transmit those signals to machines. "I knew then what I wanted to do, and I had a plan for getting there."

The first step was to drop chess and dive into an area that was attracting (and arguably, shaping) the brains of many in his generation: video games. By seventeen, he was the lead developer on the game "Theme Park." It sold millions of copies and won industry awards. He went on to Cambridge for a degree in computer science and then founded a video game company, Elixir Studios, when he was twenty-two. While running the company, Hassabis participated in the British "Mind Sports Olympiad" every year. This was where brain games aficionados gathered to compete in all kinds of contests, including chess, poker, bridge, go, and backgammon. In six years, he won the championship five times.

The way Hassabis described it, this was all leading to his current research. The video game experience gave him a grounding in software, hardware, and an understanding of how humans and computers interacted (known in the industry as man-machine interface). The computer science delivered the tools for AI. And in 2005 he went for the last missing piece, a doctorate in neuroscience.

In his current research at the Gatsby Computational Neuroscience Unit at University College, London, Hassabis focuses on the hippocampus. This is the part of the brain that consolidates memories, sifting through the torrents of thoughts, dialogues, sounds, and images pouring into our minds, and dispatches selected ones into long-term memory. Something singular occurs during that process, he believes. He thinks that it leads to the creation of concepts, a hallmark of human cognition.

"Knowledge in the brain can be separated into three levels," he said: perceptual, conceptual, and symbolic. Computers can master two of the three, perceptions and symbols. A computer with vision and audio software can easily count the number of dogs in a kennel or measure the decibel level of their barks. That's perception. And as Watson and others demonstrate, machines can associate symbols with definitions. That's their forte. Formal ontologies no doubt place "dog" in the canine group and link to dozens of subgroups, from chihuahuas to schnauzers.

But between perceiving the dog and identifying its three-letter symbol, there's a cognitive gap. Deep down, computers don't know what dogs are. They cannot create "dog" concepts. A two-year-old girl, in that sense, is far smarter. She can walk into that same kennel, see a Great Dane standing next to a toy poodle, and say, "Doggies!" Between seeing them and naming them, she has already developed a concept of them. It's remarkably subtle, and she might be hard-pressed, even as she grows older, to explain exactly how she figured out that other animals, like groundhogs or goats, didn't fit in the dog family. She just knew. Philosophers as far back as Plato have understood that concepts are essential to human thought and human society. And concepts stretch far beyond the kennel. Time, friendship, fairness, work, play, love, cruelty, peace—these are more than words. Until computers can grasp them, they will remain stunted.

The concept generator in the brain, Hassabis believes, is the hippocampal neocortex consolidation system. It has long been known that the hippocampus sifts through traces of memories, or episodes. But studies in recent years in Belgium and at MIT have probed the mechanisms involved. When rats follow a trail of food through a maze, it triggers a sequence of neurons firing in their hippocampus. Later, during the dream-filled stage known as slow-wave sleep, that same sequence is replayed repeatedly, backward and forward—and at speeds twenty times faster. Experiments on humans reveal similar patterns.

This additional speed, Hassabis believes, is critical to choosing memories and, perhaps, refining them into concepts. "This gives the high-level neocortex a tremendous number of samples to learn from," he says, "even if you experience that one important thing only once. Salient memories are biased to be replayed more often."

It isn't only that brains focus on the important stuff during dreams—a tear-filled discussion about marriage, a tense showdown with the boss. It's that they're able to race through these scenes again and again and again. It's as if TV news editors got hold of the seventeen hours of each person's waking life, promptly threw out all the boring material, and repeated the highlights ad nauseam. This isn't unlike what many experienced in the days after September 11, 2001, when the same footage of jets flying into skyscrapers was aired repeatedly. And if those images now bring to mind certain concepts, such as terrorism, perhaps it's because the hippocampus, on those late summer nights of 2001, was carrying on additional screenings, racing through them at twenty times the speed, and searing them into our long-term memories.

Even if Hassabis is right about the storage of memories and the development of concepts, transferring this process to computers is sure to be a challenge. He and his fellow researchers in London have to distill the brain's editing process into algorithms. They will instruct computers to select salient lessons, lead them to experience them repeatedly, and—it's hoped—learn from them and use them to develop concepts. This is only one of many efforts to produce a cognitive leap in computing. None of them promises rapid results. The technical obstacles are daunting, and they require the very brand of magic—breakthrough ideas—that scientists are hoping to pass on to computers. Hassabis predicts that the process will take five years. "I think we'll have something by then," he said.

8

A Season of Jitters

FROM THE VERY FIRST meeting at Culver City, back in the spring of 2007, through all the discussions about a man-machine *Jeopardy* showdown, one technical issue weighed on *Jeopardy* executives above all others: Watson's blazing speed to the buzzer. In a game featuring information hounds who knew most of the answers, the race to the signaling device was crucial. Ken Jennings had proven as much. He wouldn't have had a chance to show off his lightning fast mind without the support of his equally prodigious thumb. To Harry Friedman and his associate, the producer Rocky Schmidt, it didn't seem fair that the machine could buzz without pressing a button. They looked at it, naturally enough, from a human perspective. Precious milliseconds ticked by as the command to buzz made its way from the player's brain through the network of neurons down the arm and to the hand. At that point, if you watched the process in super slow motion, the button would appear to sink into the flesh of the thumb until—finally—the pressure triggered an electronic pulse, identical to Watson's, asking for the chance to respond to the clue. In this aspect of the game, humans were dragged down by the physical world.

It was as if they were fiddling with fax machines while Watson sent e-mails. So in a contentious conference call one morning in March 2010, the *Jeopardy* contingent laid down the law: To play against humans, Watson would also have to press the button. The computer would need a finger.

Later that day, a visibly perturbed David Ferrucci arrived late for lunch at an Italian restaurant, Il Tramonto, just down the hill from the Hawthorne labs. He joined Watson's hardware chief, Eddie Epstein, and J. Michael Loughran, the press officer who had played a major role in negotiating the *Jeopardy* challenge. Ferrucci insisted that he understood the logic behind the demand for a new appendage. And he knew that if his machine benefited from what appeared to be an unfair advantage, any victory would be tainted. What bugged him was that the *Jeopardy* team could shift the terms of the match as they saw fit, and at such a late hour.

Where would it stop? If IBM's engineers fashioned a mechanical finger that worked at ten times the speed of a human digit, would *Jeopardy* ask them to slow it down? Ferrucci didn't think so. But it was a concern. "There are deep philosophical issues in all of this," he said. "They're getting in there and deciding to graft human limitations onto the machine in order to balance things."

While the two companies shared the same broad goals, they addressed different constituencies and had different jewels to protect. If Harry Friedman and company focused first on great entertainment, Ferrucci worried, they might tinker with the rules through the rest of the year, making adjustments as one side or the other, either human or machine, appeared to gain a decisive edge. In that case, the basis for the science of the *Jeopardy* challenge was out the window. Science demanded consistent, verifiable data, all of it produced under rigorous and unchanging conditions. For IBM researchers to publish academic papers on Watson as a specimen of Q-A, they would need such data. For Ferrucci's team, building the machine alone was a career-making opportunity. But creating the scientific record around it justified the effort among their peers. This was no small consideration for a team of Ph.D.s, especially on a project whose promotional pizzazz raised suspicion, and even resentment, in the computer science community.

In these early months of 2010, tension between the two companies, and between the dictates of entertainment and those of science, was ratcheting up. As the *Jeopardy* challenge started its stretch run, IBM and *Jeopardy* entered a period of fears and jitters, marked by sudden shifts in strategy, impasses, and a rising level of apprehension.

In this unusual marriage of convenience, such friction was to be expected, and it was only normal that it would be coming to the surface at this late juncture. For two years, both *Jeopardy* and IBM had put aside many of the most contentious issues. Why bother hammering out the hard stuff—the details and conditions of the match and the surrounding media storm—when it was no sure thing that an IBM machine would ever be ready to play the game?

That was then. Now the computer in question, the speedy version of Watson, was up the road in Yorktown thrashing humans on a weekly basis. The day before the finger conversation, it had won four of six matches and put up a good fight in the other two. Watson, while still laughably oblivious in entire categories, was emerging as a viable player. The match, which had long seemed speculative, was developing momentum. A long-gestating cover story on the machine in the *New York Times Magazine* would be out in the next month or

so. Watson's turn on television was going to take place unless someone called a halt. IBM certainly wasn't about to. But
Jeopardy was another matter. *Jeopardy*'s executives now had to
consider how the game might play on TV. They had to envi-
sion worst-case scenarios and what impact they might have on
their golden franchise. As they saw it, they had to take steps to
protect themselves. Adding the finger was just one example. It
wasn't likely to be the last.

Ferrucci ordered chicken escarole soup and a salmon pa-
nini. He had the finger on his mind. "So, they come in and
say, 'You know, we don't like how you're buzzing. We're going to give you a human hand,'" he said. "This is like going to
Brad Rutter or Ken Jennings and saying, 'We're going to cut
your hands off and give you someone else's hands.' That guy's
going to have to retrain. It's a whole new game, because now
you're going to have to be a different player. We've got to re-
tune everything. Everything changes. You want to give me another nine months? You give me nine months at this stage and
. . . I don't know if I have the stomach."

From Ferrucci's perspective, the match was intriguing precisely because the contestants were different. Each side had its
own strengths. The computer could rearrange numbers and
letters in certain puzzle clues with astonishing speed. The human understood jokes. The computer flipped through millions of possibilities in a second; the human, with a native
grasp of language, didn't need to. Trying to bring them into
synch with each other would be impossible. What's more, he
suspected that any handicapping would target only one of the
parties: his machine. Just imagine, he said, laughing, if they
decided that the humans had an unfair advantage in language.
"They could give them the clues in ones and zeros!"

Nonetheless, the *Jeopardy* crew seemed intent on balancing the two sides. Another buzzer issue had come up earlier in the month. In order to keep players from buzzing too quickly, before the light came on, *Jeopardy* had long instituted a quarter-second penalty for early buzzers. The player's buzzer remained locked out during that period—a relative eternity in *Jeopardy*—and gave other, more patient rivals a first crack at the clue. But Watson, whose response was activated by the light, never fell into that trap. Its entire *Jeopardy* existence was engineered to be penalty free. So shouldn't *Jeopardy* remove the penalty for the human players as well?

For Ferrucci, this change spelled potential disaster. Humans could already beat Watson to the buzzer by anticipating the light, he said. Jennings was a master at it, and plenty of humans in sparring sessions had proven that Watson, while fast, was beatable. The electrical journey from brain to finger took humans two hundred milliseconds, about ten times as long as Watson. But by anticipating, many humans in the sparring sessions had buzzed within single milliseconds of the light. Greg Lindsay had demonstrated the technique in the three consecutive sparring sessions he'd won. If *Jeopardy* lifted the quarter-second penalty, humans could buzz repeatedly as early as they wanted while Watson waited for the light to come on. Picture a street corner in Manhattan where one tourist waits obediently for the traffic light to change while masses of New Yorkers blithely jaywalk, barely looking left or right. In a *Jeopardy* game without a penalty for early buzzing, Watson might similarly find itself waiting at the corner—and lose every buzz.

The IBM researchers could, of course, teach Watson to anticipate the buzz. But it would be a monumental task. It might require outfitting Watson with ears. Then they'd have to study the patterns of Alex Trebek's voice, the time it took him to read clues of differing lengths, the average gap in milliseconds between his last syllable and the activation of the light. It would require the efforts of an entire team and exhaustive testing during the remaining sparring sessions, made more difficult because Trebek, raised in Canada, had different voice patterns than his IBM fill-in, Todd Crain, from Illinois. It would amount to an entire research project—which would likely be useless to IBM outside the narrow confines of a specific game show. Ferrucci wouldn't even consider it.

Loughran thought Ferrucci and Friedman could iron out many of these points with a one-on-one conversation. "Why don't you pick up the phone and call Harry?" he said. "You negotiate. If they get the finger, you get rid of the anticipatory buzzing."

Ferrucci shrugged. His worries ran deeper than the finger and the buzzer. He was far more concerned about the clues Watson would face. Unlike chess, *Jeopardy* was a game created, week by week, by humans. A team of ten writers formulated the clues and the categories. If they knew that their clues would be used in the man-machine match, mightn't they be tempted, perhaps unconsciously, to test the machine? "As soon as you create a situation in which the human writer, the person casting the questions, knows there's a computer behind the curtain, it's all over. It's not *Jeopardy* anymore," Ferrucci said. Instead of a game for humans in which a computer participates, it's a test of the computer's mastery of human skills. Would a pun trip up the computer? How about a phrase in French? "Then it's a Turing test," he said. "We're not doing the Turing test!"

To be fair, the *Jeopardy* executives understood this issue and were committed to avoiding the problem. The writers would be kept in the dark. They wouldn't know which of their clues and categories would be used in the Watson showdown. According to the preliminary plans, they would be writing clues for fifteen Tournament of Champions matches, and Watson would be playing only one of them. But Ferrucci didn't think this was sufficient. One way or another they would be influenced by it, or at least they *might* be. From a scientific standpoint, there was no distinction between the existence and the possibility of bias. Either way, the results were compromised. Fifteen games, he said, was not a big enough set. "That's not statistically significant."

Epstein said that claims of bias always came up in man-machine contests, because humans always changed their behavior when faced with a machine while other humans were busy tweaking the machine. "Even in the Deep Blue chess game," he said, "Kasparov was complaining bitterly that the IBM team cheated." But how could a machine cheat in chess? "Nobody's writing questions," he said.

The concern in the chess match, Ferrucci said, was that the humans responded to Kasparov's tactics and retuned the computer. Kasparov had already adjusted to the computer's strategy and then found himself facing another one. "He was very offended by that," Ferrucci said.

"So it was unfair for the machine to change its strategy," Epstein asked, "but OK for the man to change his?"

Throughout the meal, they discussed the nature of competitions between people and machines. They weren't new, by any stretch. But earlier in the process, they had seemed more theoretical. Now, with *Jeopardy* laying down the law, theory was colliding with reality.

"I have a question for you," Epstein said at one point. "Has anyone discussed what risks *Jeopardy* has in this?"

"It raises interesting issues," Ferrucci said. "One of them

is, do they have a horse in the race? Do they want something in particular to happen? We don't control anything but our machine," he went on. "We want our machine to win. This is not a mystery. *Jeopardy* holds a different set of cards."

"They want it to be entertaining," Loughran said.

"But what does it mean for the show for the computer to win or lose?" Ferrucci asked. "What does it mean for the show if the human, let's say, clobbers the computer? These are open questions. They're in a tough spot, because on the one hand they have to maintain the [show's] integrity. But at the same time, there's a perception issue, and people might think: 'Gee, would *Jeopardy* be obsolete if the computer won? Would this change the game?'"

"No way," Loughran said.

"You don't think so," Ferrucci said, "but they have to be asking the question." He paused and ate quietly for a few moments. This marketing side of the project, which made it so exciting, was also causing stress. He was spending more and more time dealing with the *Jeopardy* team and the PR machine and less time in the lab. He was having trouble sleeping. He turned back to Loughran. "So," he asked, "knowing everything you know now, would you still do this project?"

"Sure," Loughran said. "And you?"

"I'm a science guy, so I absolutely would," Ferrucci said. He had been able to build his machine, after all, despite his concerns about how the *Jeopardy* match would play out. "But if I was a marketing guy," he added, "I'm not so sure . . ."

"We've got some issues, but it's fun," Loughran said. "We'll get through it all."

In the following days, Ferrucci looked to buffer the science of the *Jeopardy* challenge from the intrusions of the marketing effort and from the carnival odds of a one-game showdown. He devised a two-track approach for Watson, one for the scientific record, the other for the show biz extravaganza. What he wanted, he said, was a set of sixty sparring rounds in the fall of 2010 with the top *Jeopardy* players—Tournament of Champions qualifiers. These test games would be played on boards written for humans. There would be no bias toward the machine, unconscious or not. Watson would win some of the matches and lose others. But those games would represent its record against a high level of competition. It would establish a benchmark for Q-A technology and produce a valuable set of data. Even if Watson went on to stumble on national television, its reputation among the tech and scientific communities would be assured. "Those games will be where we'll get the real statistics on how we did," he said. "The final game is fun. But these sixty matches will be the real study."

Through the month of April, on conference calls and in meetings, Ferrucci repeatedly voiced his concerns to the *Jeopardy* team. He wasn't concentrating on the finger anymore. He had made that concession, and a hardware team at IBM was busy creating one. They estimated that it would slow Watson's response time by eight milliseconds. But Ferrucci continued to push for the sixty matches with champions. In April, *Jeopardy*'s Friedman and Schmidt came to watch a sparring match. In the meeting with them that followed, Ferrucci went on at length about unconscious writers' bias and tainted questions. "Dave really hammered on these points," said one participant. The *Jeopardy* executives defended their processes and protocols. The conversation grew heated. A camera crew was filming the meeting for a documentary. They were asked to leave.

That was when *Jeopardy*, in Friedman's term, "stepped back." In late April, Friedman's team sent word to IBM that they were reconsidering every aspect of the competition, including the match itself. With this news, Watson was suddenly put into the same powerless position as thousands of other *Jeopardy* wannabes: waiting for an invitation. Unlike the aspiring human players, though, Watson had no other occupation, no other purpose on earth. What's more, it had the hopes of a $96 billion corporation resting on it. And within weeks, millions of *New York Times* readers would be learning about the coming match in a Sunday magazine cover story—unless Loughran, IBM's press officer, alerted the *Times* that the match was in trouble. He keep quiet, trusting that the two sides would resolve their disagreements.

A week later, Friedman was sitting in his office on the Sony lot in Culver City. The walls were plastered with photographs and awards from his forty-year career in game shows, his seven Emmys, and his Cable and Broadcasting Hall of Fame plaque. It had been a tense day. That morning he had had another contentious phone conversation with Ferrucci, according to IBM. And he had to iron out strategy with Rocky Schmidt and Lisa Broffman, another producer on the show, before Schmidt flew to Europe the next day. "We've been so immersed in this," Friedman said, minutes after meeting with Schmidt, "that we're stepping back just a little bit and thinking of the various ramifications. We're analyzing every aspect now. This is a big deal."

Ferrucci's concerns about bias left the *Jeopardy* executives feeling exposed. The IBM scientist, after all, was implying that *Jeopardy*'s writers might tilt the match toward one side or the other—or at least be perceived as doing so. Ferrucci was always careful to ascribe this possibility to unconscious bias. But for *Jeopardy*, a franchise born from the quiz show scandals of the 1950s, the hint of such bias—conscious or not—was poisonous. And even if Ferrucci kept this concern to himself, the point he made repeatedly was that other scientists would raise the very same questions. If it was even within the realm of possibility that *Jeopardy* had an interest in the outcome and if it used its own people to write the clues, the fairness of the game and the validity of the contest were compromised.

For Friedman, who took pride in lending the *Jeopardy* platform to science, this was tough to swallow. "[IBM] could have done this with a bunch of questions that academics came up with," he said. "But they wanted this fabulous platform. They gain the platform and lose control." He maintained that the future of the franchise hinged on its reputation for fairness and integrity and that if the match went forward, his team would be laying down the rules. "We rigidly adhere to not only our own code of conduct, but also obviously to the FCC regulations," he said. "We run a pretty tight ship."

He described how the contestants are sequestered during the filming, accompanied by handlers and prohibited from mingling with anyone with access to the clues. He recalled one time that Ken Jennings, hurrying to change a tie that "strobed on camera," ducked into a little nook where Alex Trebek checked his appearance before stepping onto the set. This was a breach. The three players had to always stick together, under surveillance, so that no one could even be suspected of receiving favorable treatment. Jennings was quickly ousted as if he'd been a North Korean commander strolling into a meeting of the Joint Chiefs at the Pentagon. Friedman laughed. "He could have been shot." Then he play-acted. "Oh, sorry Ken, we had to wing you in your foot there, but your buzzer thumb seems to be intact. Are you OK to play the next show? You wandered into a secure area . . ."

Friedman brushed off Ferrucci's suggestion that the results

of the game could have a lasting impact on the *Jeopardy* franchise, much as Kasparov's loss to Deep Blue forever changed chess. He laughed. "When all of this, as wonderful as it is, is over, we're going to continue playing our game. We're going to continue what got us here through six thousand shows." The message to IBM: "Thanks for coming. Thanks for playing. We're back to our day jobs."

The tentative plan had been for the IBM team to move Watson to the Culver City studios in late 2010. It would participate in a championship match, playing against Ken Jennings and the winner of an invitational tournament of past champions. But bringing the machine into *Jeopardy*'s "tightly run ship," it was now clear, raised complications, including demands to change the show's tried-and-tested procedures. It raised the risk of rancor and public accusations. And it wasn't just the scientists who might complain. The humans would be playing for a million-dollar prize, underwritten by IBM. If they suspected any tilting in the competition, they were sure to speak up as well. In a sense, Watson's intrusion into the *Jeopardy* world represented a potential breach of its own. Friedman had to weigh his options.

One of *Jeopardy*'s biggest fears, Ferrucci believed, was that Watson would grow dramatically smarter and faster over the summer and lay waste to its human foes. This was early May, weeks after *Jeopardy* had begun to reconsider the match. He was sitting in the empty observation room on the *Jeopardy* set in Yorktown. At the podium on the other side of the window, Watson had been beating humans in sparring sessions about 65 percent of the time but showing few signs of frightening dominance. The *Jeopardy* crew, he said, continued to assess the matches. "Is this fun, is this entertaining, is this speaking to our audience?" A superendowed Watson, conceivably, would drain the match of all suspense. In that case, according to Ferrucci, "People would say, 'Of course computers can beat humans! Why did you promote all this?'"

Ferrucci wished it were true, that with a few devilishly smart new algorithms Watson would leap forward into a class of its own. That way he might sleep better. But he didn't see it happening. "We're working our butts off," he said. "But I don't think we're going to see a lot of difference in Watson's performance four months from now, when we have to freeze the system. But *they* don't know that," he said. "How could they know? They're not doing the science."

Jeopardy's executives also worried, he said, that IBM could jack up Watson's speed simply by adding more computing power. This was logical. But it was not the case. In distributing Watson's work to more than two thousand processors, the IBM team had broken it into hundreds of smaller tasks, most of them operating in parallel. But a handful of these jobs, Ferrucci explained, required sequential analysis. Whether it was parsing a sentence or developing a confidence ranking for a potential answer, certain basic algorithms had to follow strings of commands, with each step hinging on the previous one. This took time.

Think of a billionaire selecting his outfit for a black-tie event. He can assign some tasks to his minions. One can buy socks while others track down shoes, pants, and a shirt. Those jobs, in computer lingo, run in parallel. But when it comes to getting dressed, the work becomes sequential. The man must place one leg in his pants, then the other. Maybe a few butlers could help with his socks simultaneously and hold out the arms of his shirt for him, but such opportunities are limited. This sequence, to the last snap of the cuff links, takes time.

Inside Watson, some of the sequential algorithms gobbled up a quarter of a second, half a second, even more. And they could not be shared among many machines. Watson, in all likelihood, would need the same two to five seconds by the date of the final match. At this point, the only path to greater speed was to come up with simpler commands—smarter algorithms that led Watson through fewer steps. But Ferrucci didn't expect advances of more than a few milliseconds in the coming months. Nonetheless, he found it hard to make his case to the *Jeopardy* team. From their perspective, Watson had risen from a slow-witted assortment of software into a champion-caliber player in two years. Who was to say it wouldn't keep improving?

In this jittery home stretch, it was becoming clear, the two sides shared parallel fears. While Hollywood worried that the computer would grow too smart, the IBM team focused on its vulnerabilities and fretted that it would fail. Watson's weekly blunders in the sparring sessions added to the long lists of bugs to eliminate, mauled pronunciations to remedy, potential gaffes to program around. There wasn't enough time to address them all. In the same pragmatic spirit that had marked the entire enterprise, they carried out time-benefit analyses on their list of items and focused on the ones at the top. "This is triage," said Jennifer Chu-Carroll.

One small but vital job was to equip Watson with a profanity filter. The machine had already demonstrated, by dropping the F-bomb on its answer panel, how heedless it could be to basic norms of etiquette and decency. The simplest approach would be to prohibit it from even considering the seven forbidden words that George Carlin made famous in his comedy routines, plus a handful of others, including ethnic and racial slurs. It would be easy to draw up a set of rules—heuristics—to override the machine's statistically generated candidate answers. But what about words that included no-no's? Consider this 2006 clue in the category T Birds: "In North America this term is properly applied to only 4 species that are crested, including the tufted." Would a list of forbidden vulgarities impede Watson from answering, "What is a titmouse?" Researchers, said David Gondek, would have to come up with "loose filters," leaving room for such exceptions. But they were sure to miss some.

Then there was the matter of pronunciation. Watson could turn an everyday word into a profanity with just a slip of its mathematically programmed tongue. This was even more likely with foreign words. How would it fare, for example, answering this 2007 clue in the Plane Crazy category? "In 1912 this Dutch plane builder set up a plant near Berlin; later, his fighter planes were flown by the Red Baron." This would likely be a slam-dunk for Watson, but leading it to correctly enunciate "What is Fokker?" would involve meticulous calibration of its vowel pronunciation. Surely, some would say, *Jeopardy* would not include a Fokker clue in a match involving a machine. But that would revive Ferrucci's key concern: that *Jeopardy* would be customizing the game for Watson. In the end, Watson's scientists could only fashion a profanity filter, make room for the most common exceptions, tweak potentially problematic pronunciations, and hope for the best. If the machine, despite their work, found a way to say something outrageous, it would be up to the show's producers to bleep it out.

While her colleagues steered Watson away from gaffes, Chu-Carroll was concentrating on Final Jeopardy, an area of mounting concern for Ferrucci's team. Final Jeopardy was often decisive. Throughout Watson's training, the team had studied and modeled all of the clues as a single group. They knew from the beginning that the Final Jeopardy clues were trickier—"less direct, more implicit," in Chu-Carroll's words—but their data set of these clues was much smaller, only one sixty-first of the total. Because of this, the computer was still treating the Final Jeopardy clue like every other clue on the board, coming up with its answer in three to five seconds—and then just waiting as the thirty-second jingle went through its sixty-four notes. This was enough time for trillions of additional calculations. Wasn't there a way to take advantage of the extra seconds?

The team was not about to devise new ways to find answers. That would require major research. But Watson could take more time to analyze the answers it collected. The method, like most of Watson's cognitive work, would require exhaustive and repetitive computing. The idea was to generate from each answer a series of declarative statements, then check to see if they looked right. In the category English Poets, for example, one recent Final Jeopardy clue had read: "Translator Edward Fitzgerald wrote that her 1861 'death is rather a relief to me . . . no more Aurora Leighs, thank God.'" Let's say Watson came up with measurable confidence in three potential names, Alfred Lord Tennyson, Emily Dickinson, and Elizabeth Barrett Browning. It could then proceed to craft statements, putting each name in the following sentences: "_____ died in 1861," "_____ wrote Aurora Leigh," "_____ was an English poet." Naturally, some of the sentences would turn out to be foolish, perhaps: "_____ found relief in death" or "_____ died, thank God." In any case, for each of dozens of sentences, Watson would race through its database looking for matches. This represented an immense amount of work. But the results could boost its confidence in the correct response—"Who is Elizabeth Barrett Browning?"—and guide it toward acing Final Jeopardy.

James Fan, meanwhile, was going over clues in which Watson failed to understand the subject. At one meeting at the Hawthorne labs, he brought up an especially puzzling one. In the category Postal Matters, it asked: "The first known air mail service took place in Paris in 1870 by this conveyance." From its analysis, Watson could conclude that it was supposed to find a "conveyance." That was the lexical answer type, or LAT. But what was a conveyance? In all of the ontologies it had on hand, there was no such grouping. There were groups of trees, colors, presidents, even flavors of ice cream—but no "conveyances." And if Watson looked up the word, it would find vague references to everything from communication to the transfer of legal documents. One of its meanings involved transport, but the computer would hardly know to focus its search there.

What to do? Fan was experimenting with a new grouping of LATs. At a meeting of one algorithm team on a June afternoon, he started to explain how he could prepare Watson for what he called weird LATs.

Ferrucci didn't like the sound of it. "We don't have any way to mathematically classify 'weird,'" he objected. "That's a word you just introduced." Run-of-the-mill LATs, such as flowers, presidents, or diseases, provided Watson with vital intelligence, dramatically narrowing its search. But an amorphous grouping of "weird" words, he feared, would send the computer off in bizarre directions, looking at more distant relationships in the clue and bringing in masses of erroneous possibilities, or noise.

"There are ways to measure it," Fan said. "We can look at how many instances there are of the LAT in Yago"—a huge

semantic database with details on more than two million entities. "And if it isn't there, we can classify it as "weird.""

"Just based on frequency?" Ferrucci said. There were only
weeks left to program Watson, and he saw this "weird" grouping as a wasteful detour. In the end, he gave Fan the go-ahead.
"If something looks hare-brained and it's only going to take a
couple of days, you do it." But he worried that such last-min-
ute fixes might help Watson on a couple of clues and disorient
it on many others. And there were still so many other problems to solve.

By the end of June, two weeks after Watson graced the cover
of the *New York Times Magazine,* Harry Friedman had come
to a decision. The solution was to remove the man-machine
match, with all of its complications, from *Jeopardy*'s programming schedule. "This is an exhibition," he said, adding that
it made the "whole process a lot more streamlined." *Jeopardy*
would follow its normal schedule. The season of matches
would feature only humans. Writers would follow the standard protocols. Nothing would change. The Watson match,
with its distinct rules and procedures, would exist in a world
of its own. In a call to IBM, Friedman outlined the new rules
of engagement. The match would take place in mid-January
at IBM Research. It would feature Ken Jennings and Brad
Rutter in two half-hour games. The winner as in all *Jeopardy*
tournaments, would be the player with the highest combined
winnings.

Friedman addressed Ferrucci's concerns about writers' bias
by enlarging the pool of games. Each year the *Jeopardy* writers produced about a hundred games for the upcoming season, with taping starting in July. A few days before taping, an
official from Sullivan Compliance Company, an outside firm

that monitors game shows, would select thirty of those games. He would not see the clues or categories and would pick two of the games only by numbers given to them. Once the games were selected, a *Jeopardy* producer would look at the clues and categories. If any of them overlapped with those that Jennings or Rutter had previously faced, or included the types of audio and visual clues that were off-limits for Watson, the category would be removed and replaced by a similar one from another of the thirty games. If a Melville category recalled one that Jennings had faced in his streak, they might replace it with another featuring Balzac or Whitman. And for Watson's scientific demonstration, the machine would play fifty-six matches throughout the fall against Tournament of Champions qualifiers. This was the best test stock *Jeopardy* had to offer—the closest it could come to the two superstars Watson would face in January.

Jeopardy, eager for a blockbuster, had come up with a scheme to manage the risks. After months of fretting, the game was on.

9

Watson Looks for Work

DINNER WAS OVER at the Ferrucci household. It was a crisp evening in Yorktown Heights, a New York suburb ten miles north of the IBM labs. It was dark already, and the fireplace leapt with gas-fed flames. Ferrucci's two daughters were heading upstairs to bed. In the living room, Ferrucci and his wife, Elizabeth, recounted a deeply frustrating medical journey—one that a retrained *Jeopardy* computer (Dr. Watson) could have made much easier.

Ferrucci had been envisioning a role for computers in doctors' offices since his days as a pre-med student at Manhattan College. In graduate school, he went so far as to build a medical expert system to provide advice and answer questions about cardiac and respiratory ills. It worked well, he said, but its information was limited to what he taught it. A more valuable medical aid, he said, would scoop up information from anywhere and come up with ideas and connections that no one had even thought to consider. That was the kind of machine he himself had needed.

Early in the *Jeopardy* project, Ferrucci said, not long after the bake-off, he started to experience strange symptoms.

The skin on one side of his face tingled. A couple of his fingers kept going to sleep. And then, one day, searing pain shot through his head. It lasted for about twenty seconds. Its apparent epicenter was a lower molar on the right side of his mouth. "It felt like someone was driving an ice pick in there," he said.

When this pain returned, and then came back a third and fourth time, Ferrucci went to his dentist. Probing the tooth and placing ice on it, the dentist attempted to reproduce the same fearsome effects but failed. He could do a root canal, he said, but he had no evidence that the tooth was the problem. Ferrucci then went to a neurologist, who suggested a terrifying possibility. Perhaps, he said, Ferrucci was suffering from trigeminal neuralgia, more commonly known as the suicide disease. It was a nerve disorder so painful that it was believed to drive people to kill themselves. He recommended getting the root canal. It might do the trick, he said, and save them both from the trouble of ransacking the nervous system for answers.

Ferrucci got the root canal. It did no good. The attacks continued. He went to another neurologist, who prescribed anticonvulsion pills. When he read about the medicine's side effects, Ferrucci said, "I didn't know whether to take the pills or buy a gun." He did neither and got an MRI. But putting on the helmet and placing his head in the cramped cylinder filled him with such anxiety that he had to go to another doctor for sedatives.

He had no idea what was wrong, but it wasn't getting any better. As the *Jeopardy* project moved along, Ferrucci continued to make presentations to academics, industry groups, and IBM brass. But he started to take along his number two, Eric

Brown, as a backup. "If I don't make it through the talk," he told Brown, "you just pick up where I leave off."

In time, Ferrucci started to recognize a certain feeling that preceded the attacks. He sensed it one day, braced himself against a wall, lowered his head slightly, and awaited the pain. It didn't come. He moved his head the same way the next time and again he avoided the pain. He asked his neurologist about a possible link between the movements of his neck and the facial pain. He was told there was no possible connection.

Months passed. The Ferruccis were considering every option. "Someone told us we should get a special mattress," said Elizabeth. Then a friend suggested a specialist in craniofacial pain. The visit, Ferrucci learned, was not covered by his insurance plan and would cost $600 out of pocket. He decided to spend the money. A half hour into the meeting, the doctor worked his hands to a spot just below Ferrucci's collarbone and pressed. The pain rocketed straight to the site of his deadened molar. The doctor had found the connection. A number of small contraction knots in the muscle, myofascial trigger points, created the pain, he said. The muscle was traumatized, probably due to stress. With massage, the symptoms disappeared. And Ferrucci kept them at bay by massaging his neck, chest, and shoulders with a two-dollar lacrosse ball.

He walked to a shelf by the fireplace and brought back a book, *The Trigger Point Therapy Workbook,* bristling with Post-it notes. On one page was an illustration of the sternocleidomastoid. It's the biggest visible muscle in front of the neck and extends from the sternum to a bony prominence behind the ear. According to the book, trauma within this muscle could cause pain in the head. In a single paragraph on page 53 was a description of Ferrucci's condition. It had references to toothaches in back molars and a "spillover of pain . . . which mimics trigeminal neuralgia." Ferrucci could have written it himself.

If a computer like Watson, customized for medicine, had access to that trigger point workbook along with thousands of other books and articles related to pain, it could have saved Ferrucci a molar and months of pain and confusion. Such a machine likely would have been able to suggest, with at least some degree of confidence, the connection between Ferrucci's symptoms and his sternocleidomastoid. This muscle was no more obscure than the Asian ornaments or Scandinavian kings that Watson routinely dug up for *Jeopardy*. Such a machine would not have to understand the connections it found. The strength of the diagnostic engine would not be its depth, but its range. That's where humans were weak. Each expert that Ferrucci visited had mastered a limited domain. The dentist knew teeth, the neurologists nerves. But no one person, no matter how smart or dedicated, could stay on top of discoveries across every medical field. Only a machine could do that.

A few months earlier, on an August morning, about a hundred IBM employees filed into the auditorium at the Yorktown labs. They included researchers, writers, marketers, and consulting executives. Their goal was to brainstorm ideas for putting Watson to work outside the *Jeopardy* studio. The time for games was nearly over. Watson, like thousands of other gifted students around the world, had to start earning its keep. It needed a career.

This was an unusual situation for an IBM product, and it indicated that the company had broken one of the cardinal rules of technology development. Instead of focusing first on a market opportunity and then creating the technology for it, IBM was working backward: It built the machine first and was now wondering what in the world to do with it. Other tech companies were notorious for this type of cart-before-the-horse innovation. Motorola, in the 1990s, led the development of a $5 billion satellite phone system, Iridium, before learning that the market for remote communications was tiny and that most people were satisfied with normal cell phones. Within a year of its launch, Iridium went bankrupt. In 1981, Xerox built a new computer, the 6085 Star, featuring a number of startling innovations—a mouse, an ethernet connection, e-mail, and windows that opened and closed. All of this technology would lay the groundwork for personal computers and the networked world. But it would be other companies, notably Apple and Microsoft, that would take it to market. And in 1981, Xerox couldn't find buyers for its $16,000 machines. Would Watson's industry-savvy offspring lead to similar boondoggles?

In fairness to IBM, Grand Challenges, like Watson and the Deep Blue chess machine, boosted the company's brand, even if it came up short in the marketplace. What's more, the technology developed in the *Jeopardy* project, from algorithms that calculated confidence in candidate answers to wizardry in the English language, was likely to work its way into other offerings. But the machine's question-answering potential seemed so compelling that IBM was convinced Watson could thrive in a host of new settings. It was just a question of finding them.

Ferrucci started the session by outlining Watson's skills. The machine, he said, understood questions posed in natural language and could read millions of documents and scour databases at lightning speed. Then it could come up with responses. He cautioned his colleagues not to think of these as answers but hypotheses. Why the distinction? In every domain most of Watson's candidate answers would be wrong. Just as in *Jeopardy*, it would come back with a list of possibilities. People looking to the machine for certainty would be disappointed and perhaps even view it as dumb. Hypotheses initiate a lengthier process. They open up paths of inquiry. If Watson came back from a hunt with ten hypotheses and three of them looked promising, it wouldn't matter much if the other seven were idiotic. The person using the system would focus on the value. And this is where the vision of Watson in the workplace diverged from the game-playing model. In the workplace, Watson would not be on its own. Unlike the *Jeopardy* machine, the Watson Ferrucci was describing would be engineered to supplement the human brain, not supplant it.

The time looked ripe for word-savvy information machines like Watson, thanks to the global explosion of a new type of data. If you analyzed the flow of digital data in, say, 1980, only a smidgen of the world's information had found its way into computers. Back then, the big mainframes and the new microcomputers housed business records, tax returns, real estate transactions, and mountains of scientific data. But much of the world's information existed in the form of words—conversations at the coffee shop, phone calls, books, messages scrawled on Post-its, term papers, the play-by-play of the Super Bowl, the seven o'clock news. Far more than numbers, words spelled out what humans were up to, what they were thinking, what they knew, what they wanted, whom they loved. And most of those words, and the data they contained, vanished quickly. They faded in fallible human memories, they piled up in Dumpsters and moldered in damp basements. Most of these words never reached computers, much less networks.

That has all changed. In the last decade, as billions of people have migrated their work, mail, reading, phone calls, and webs of friendships to digital networks, a giant new species of data has arisen: unstructured data. It's the growing heap of sounds and images that we produce, along with trillions of words. Chaotic by nature, it doesn't fit neatly into an Excel spreadsheet. Yet it describes the minute-by-minute goings-on of much of the planet. This gold mine is doubling in size every year. Of all the data stored in the world's computers and coursing through its networks, the vast majority is unstructured. Hewlett Packard, for example, the biggest computer company on earth, gets a hundred fifty million Web visits a month. That's nearly thirty-five hundred customers and prospects per minute. Those visits produce data. So do notes from the company's call centers, online chat rooms, blog entries, warranty claims, and user reviews. "Ninety percent of our data is unstructured," said Prasanna Dhore, HP's vice president of customer intelligence. "There's always a gap between what you want to know about the customer and what is knowable." Analysis of the pile of data helps reduce that gap, bringing the customer into sharper focus.

The potential value of this information is immense. It explains why Facebook, a company founded in 2004, could have a market value six years later of $50 billion. The company gathers data, most of it unstructured, from about half a billion people. Beyond social networks and search engines, an entire industry has sprung up to mine this data, to predict people's behavior as shoppers, drivers, workers, voters, patients, even potential terrorists. As machines, including

Watson, have begun to chomp on unstructured data, a fundamental shift is occurring. While people used to break down their information into symbols a computer could understand, computers are now doing that work by themselves. The machines are mastering human communication.

This has broad implications. Once computers can handle language, every person who can type or even speak becomes a potential programmer, a data miner, and an analyst. This is the march of technology. We used to have typists, clerks, legions of data entry experts. With the development of new tools, these jobs became obsolete. We typed (and spell-checked), laid out documents, kept digital records, and even developed our own pictures. Now, a new generation of computers can understand ordinary English, hunt down answers in vast archives of documents, analyze them, and come up with hypotheses. This has the potential to turn entire industries on their heads.

In the August meeting, Ferrucci told the audience the story of his recent medical odyssey and how a machine like Watson could have helped. Others suggested that Watson could man call centers, function as a brainy research assistant in pharmaceutical labs, or work as a whip-smart paralegal, with nearly instant recall of the precedents, both state and federal, for every case. They briefly explored the idea of Watson as a super question-answering Google. After all, it could carry out a much more detailed analysis of questions and piece together sophisticated responses. But this idea went nowhere. IBM had no experience in the commercial Web or with advertisers. Perhaps most important, Watson was engineered to handle one *Jeopardy* clue at a time. In those same three seconds, a search engine like Google's or Microsoft's Bing handled millions of queries. To even think about competing, the IBM team would

have to build an entirely new and hugely expensive computing architecture. It was out of the question.

No, Watson's future was as an IBM consulting tool and
there were plenty of rich markets to explore. But before Watson could make a go of it, Big Blue would have to resolve serious questions. First, how much work and expense would it
take to adapt Watson to another profession, to curate a new
body of data and to educate the machine in each domain? No
one could say until they tried. Second, and just as important,
how much resistance would these new knowledge engines encounter? New machines, after all, are in the business of replacing people—not something that often generates a warm
welcome. The third issue involved competition. Assuming
that natural-language, data-snarfing, hypothesis-spouting machines made it into offices and laboratories, who was to say
that they'd be the kin of a *Jeopardy* contraption? Other companies, from Google to Silicon Valley startups, were sure to be
competing in the same market. The potential for these digital oracles was nearly limitless. But in each industry they faced
obstacles, some of them considerable.

Medicine was one of the most promising areas but also
among the toughest to crack. The natural job for Watson
would be as a diagnostic aid, taking down the symptoms in
cases like Ferrucci's and producing lists of possible conditions, along with recommended treatments. Already, many
doctors facing puzzling symptoms were consulting software
tools known as medical decision trees, which guided them toward the most likely diagnoses and recommended treatments.
Some were available as applications on smart phones. A medical Watson, though, would plunge into a much deeper pool
of data, much of it unstructured. Conceivably, it would come
up with hidden linkages. But even that job, according to Rob-

ert Wachter, the chief of hospital medicine at the University of California, San Francisco, was bound to raise serious questions. "Doctors like the idea of having information available," he said. "Where things get more psychologically fraught is when a damned machine tells them what to do." What's more, once analysis is automated, he said, the recommendation algorithm is likely to include business analysis. In other words, the medical Watsons might come back not with the statistically most effective treatment but the most *cost-effective* one. Even if this didn't happen, many would remain suspicious. And what if Watson had sky-high confidence in a certain diagnosis—say, 97 percent? Would doctors get in trouble if they turned a deaf ear to it? Would they face lawsuits if they ignored the advice and it later turned out the machine was right?

Then, of course, there was the possibility of disastrous mistakes resulting from a computer's suggestions. Even if a bionic assistant scrupulously labeled all of its findings as hypotheses, some of them—just like Watson's answers in *Jeopardy*—were bound to be nutty, generating ridicule and distrust. Others, perhaps more dangerous, would be wrong while appearing plausible. If a treatment recommended by a machine killed a patient, confidence in bionic assistants could plummet.

The other issue, sure to come up in many industries, boils down to a struggle for power, and even survival, in the workplace. "As every profession embraces systems that take humans out of it," Wachter said, "the profession gets commoditized." He noted the example of commercial aviation, where pilots who were once considered stars have ended up spending much of the time in flight simply backing up the machines that are actually flying the planes. The result? "Pilots' pensions have been cut and they're paid less, because they're largely interchangeable," he said. "Doctors don't want to see that happening to them."

For IBM, this very scenario promises growth. With more than $4 billion in annual revenue, the health care practice within IBM Global Services has the size of a Fortune 500 company. It runs large data centers for hospitals and insurance companies. It also helps them analyze the data, looking for patterns of symptoms, treatments, and diseases—as well as ways to cut costs. This is part of a trend toward statistical analysis in the industry and the rapid growth of so-called evidence-based medicine. But one of the most valuable streams of data—the doctor's notes—rarely makes it into the picture, said Joseph Jasinski, who heads research for IBM's health care division. This is where the doctor writes down what he or she sees and thinks. Sometimes it is stored in a computer, but only, Jasinski said, "as a blob of text." In other words, it's unstructured data, Watson's forte. "There's a strong belief in the community that if you could study clinical notes, you could analyze patient similarities," he said. Neurologists' notes—going back to Ferrucci's case—could have pointed to common symptoms between patients with the suicide disease and others with knots in a muscle just below their shoulder blade. This analysis could expand, comparing symptoms and treatments, and later study the outcomes. What works? What falls flat? Which procedures appear to waste money?

Despite the growth of evidence-based medicine, many of these studies are hard to carry out, especially in the splintered $2.3 trillion American system. The doctor prescribes the treatment and the insurance company pays for it, but all too often neither of them gets the most vital feedback: how it worked. The feedback loop, in the language of statisticians, rarely closes. The most promising sites for this type of analysis, Jasinski said, are self-contained hospital networks, which keep voluminous records on patients and do extensive follow-up. They would include the Veterans' Administration, the Mayo Clinic, and Kaiser Permanente in the United States. Many countries with national health care systems also have promising data. Denmark, where IBM has been running the data since 2006, could provide a national laboratory. There, a medical Watson could diagnose diseases, suggest treatments that have proven successful, and steer doctors away from those that have led to problems. Such analyses could save lives, Jasinski said. "We kill a hundred thousand people a year from preventable medical errors."

In fact, the potential for next-generation computers in medicine stretches much further. Within a decade, it should cost less than $100 to have an individual's entire genome sequenced. Some people will volunteer to have this done. (Already, companies like 23andMe, a Silicon Valley startup, charge people $429 for a basic decoding.) Others, perhaps, will find themselves pressed, or even compelled, by governments or insurers, to submit their saliva samples. In either case, computers will be studying, correlating, and answering questions about growing collections of this biological information.

At the same time, we're surrounding ourselves with sensors that provide streams of data about our activities. Coronary patients wear blood pressure monitors. Athletes in endurance sports cover themselves with electronics that produce torrents of personal data, reading everything from calorie burn to galvanic skin response, which is associated with stress. Meanwhile, companies are rushing into the market for personal monitoring. Zeo, a Rhode Island company, sells a bedside device that provides a detailed readout every morning of a person's sleeping patterns, including rapid-eye movement, deep sleep, and even trips to the bathroom. Intel is outfitting the homes of elderly test subjects with sensors to measure practically every activity possible, from their nocturnal trips to the bathroom to the words they type on their computers. And each person who carries a cell phone unwittingly provides detailed information on his or her daily movements and routines—behavioral data that could prove priceless to medical researchers. Even if some of this data is shielded by privacy rules and withheld from the medical industry, much of it will be available. Machines like Watson will be awash in new and rising rivers of data.

But in the autumn of 2010, as Watson prepared for its culminating *Jeopardy* match, it had yet to land its first hospital job, and its medical abilities remained largely speculative. "We have to be cautious here," Jasinski said. Though full of potential, Watson was still untested.

It may seem frivolous for the IBM team to have worked as hard as it did to cut down Watson's response time from nearly two hours to three seconds. All of that engineering, and those thousands of processors were harnessed, just to be able to beat humans to a buzzer in a quiz show. Yet as Watson casts about for work, speed will be a crucial factor. Often it takes a company a day or two to make sense of the data it collects. It can seem remarkable, because the data provides a view of sales or operations that was unthinkable even a decade ago. But still, the delay means that today doesn't come into focus until tomorrow or next week. The goal for many businesses now is to process and respond to data in real time—in the crucial seconds that a quick investment could net $10 million or the right treatment could save a patient's life. Chris Bailey, director of the Advanced Computing Lab at SAS, a major producer of analytics software, says the focus is on speed. "Our goal is to make the systems run a thousand or a million times faster," he said. "That enables us to look at a million times more input." With this speed, companies increasingly will be able to carry out research, and even run simulations, while the customer is paying for a prescription or withdrawing funds.

Computers with speed and natural language are poised to transform business processes, perhaps entire industries. Compared to what's ahead, even today's state of the art looks sluggish. Consider this snapshot of the data economy, circa 2011: A man walks into a pharmacy to renew his blood pressure medication. He picks up some toiletries while he's there. He hands the cashier his customer loyalty card, which lowers his bill by a dollar or two, and then pays with his Visa card. This shopping data goes straight to Catalina Marketing in St. Petersburg, Florida, which follows the purchases of 190 million shoppers in America. Catalina scans the long list of items that this consumer has bought in the last three years and compares his patterns with those of millions of others. While he's standing at the register, it calculates the items most likely to interest him. Bundled with the receipt the cashier hands him, he finds several coupons—maybe one for oatmeal, another for a toothpaste in a new upside-down dispenser. If and when he uses them, Catalina learns more about him and targets him with ever-greater precision.

That might sound like a highly sophisticated process. But take a look at how Catalina operates, and you'll see it involves a painfully slow roundtrip, from words to numbers and then back again. "Let's say Kraft Foods has a new mac and cheese with pimentos," said Eric Williams, Catalina's chief technology officer. The goal is to come up with a target group of po-

tential macaroni eaters, perhaps a million or two, and develop
the campaign most likely to appeal to them. The marketers
cannot summon this intelligence from their computers. They
hand the instructions—the idea—to Catalina's team of seventy statisticians. For perhaps a week, these experts hunt for
macaroni prospects in the data. Eventually they produce lists,
clusters, and correlations within their target market. But their
statistical report is not even close to a marketing campaign.
For that, the marketers must translate the statisticians' results
back into words and ideas. "Trying to interpret what these
people find into common language is quite a feat," Williams
said. Eventually, a campaign takes shape. Catalina concocts
about six hundred to eight hundred of them a year. They're
effective, often doubling or tripling customer response rates.
But each campaign, on average, gobbles up a month of a stat-
istician's work.

Williams's fantasy is to have a new type of computer in his
office. Instead of delivering Kraft's order to his statisticians, he
would simply explain the goals, in English, to the machine. It
would pile through mountains of data in a matter of seconds
and come back with details about potential macaroni buyers.
The language-savvy machine wouldn't limit its search to traditional data, the neatly organized numbers featuring purchases,
dates, and product codes. It might read Twitter or scan social
networks to see what people are writing about their appetites
and dinner plans. After this analysis, Williams's dream machine could return with a list of ten recent marketing campaigns that have proven the most effective with the target
group. "If I don't have to go to statisticians and wait while
they run the data, that would be huge," Williams said. Instead
of eight hundred campaigns, Catalina might be able to handle
eighty thousand, or even a million—and offer them at a frac-

tion of today's cost. "You're talking about turning marketing on its head."

His dream machine, of course, sounds like a version of Watson. Its great potential, in marketing and elsewhere, comes from its ability to automate analysis—to take people, with their time-consuming lunch breaks and vacations, their disagreements and discussions, and drive them right out of the business. The crucial advantage is that Watson—and machines like it—eliminate the detour into the world of numbers. They understand and analyze words. Machines like this—speedy language whizzes—will open many doors for business. The question for IBM is what Watson's place will be in this drama, assuming it has one.

In the near term, Watson's job prospects are likely to be in call centers. Enhanced with voice recognition software and trained in specific products and services, the computer could respond to phone calls and answer questions. But more challenging jobs, such as bionic marketing consulting, are further off. For each industry, researchers working with consulting teams will have to outfit Watson with an entirely new set of data and run through batteries of tests and training sets. They'll have to fine-tune the machine's judgment—the degree of confidence it generates for each response—and adapt hardware to the job. Will customers want access to mini-Watsons on site or perhaps gain access to a bigger one through a distant data center, a so-called cloud-based service? At this point, no one can say. Jurij Paraszczak, director of Industry Solutions and Emerging Geographies at IBM Research, sees versions of Watson eventually fitting into a number of industries. But such work is hardly around the corner. "Watson's such a baby," he said.

• • •

The history of innovation is littered with technologies that failed because of bad timing or rotten luck. If that $16,000 Xerox computer, with the e-mail and the mouse, had hit the market a decade later, in 1991, cheaper components would have lowered the cost by a factor of five or ten and a more informed public might have appreciated its features.

Technology breakthroughs can also consign even the most brilliant and ambitious projects to museum pieces or junk. In 1825, the first load of cargo floated from Buffalo to Albany on the new Erie Canal. This major engineering work, the most ambitious to date in the Americas, connected the farms and nascent industries of the Great Lakes to the Hudson River, and on to the Atlantic Ocean. It positioned New York State as a vital thoroughfare for commerce, and New York City as the nation's premier port. The news could hardly have been worse for the business and government leaders in New York's neighbor to the south, Pennsylvania, and its international port, Philadelphia. They'd been outmaneuvered. The only way to haul cargo across Pennsylvania was by Conestoga wagon, which often took up to three weeks. So that very year, the Pennsylvanians laid plans to build their own waterway connecting Philadelphia to the Great Lakes. They budgeted $10 million, an immense sum at the time and $3 million more than the cost of the Erie Canal.

The Pennsylvania Canal faced one imposing obstacle: the Alleghenies. These so-called mountains were smaller and rounder than the Rockies or Alps, but they posed a vertical challenge for canal designers. Somehow the boats would have to cross from one side to the other. So with engineering far more ambitious than anything the New Yorkers had attempted, the Pennsylvanians constructed a complex series of iron-railed ramps and tunnels. This horse-powered system would hoist the boats back and forth over the hills and, in a few places, through them. By 1834, boats crossing the state left the water for a thirty-six-mile cross-country trek, or portage, which lifted them up fourteen thousand feet and back down again. For the criss-crossing on ramps, the boats had to be coupled and uncoupled thirty-three times. For those of us not accustomed to Conestoga wagons, this would seem excruciatingly slow, even after stationary steam engines replaced the horses. The novelist Charles Dickens made the crossing in 1842. He set off on a boat and found himself, in short order, peering down a mountain cliff. "Occasionally the rails are laid upon the extreme verge of the giddy precipice and looking down from the carriage window," he wrote, "the traveler gazes sheer down without a stone or scrap of fence between into the mountain depths below." Dickens wouldn't soon forget Pennsylvania's $10 million innovation.

This canal project used current technology and heroic engineering to carry out work on a tight deadline. Would newer technology supplant it? It was always possible. After all, the Industrial Revolution was raging in Britain and making inroads into America. Things were changing quickly, which was precisely why Pennsylvania could not afford to wait. As it turned out, within a decade steam-powered trains rendered the canal obsolete. Yet the project laid out the vision and preliminary engineering for the next generation of transport. The train route over the Alleghenies, which followed much of the same path, was considered so vital to national security that Union soldiers kept guard over it during the Civil War. And in 1942, as part of the failed Operation Pastorius, Nazi agents targeted it for sabotage.

Will Watson be the platform for the next stage of computing or, like the Pennsylvania Canal, a bold and startling curiosity to be picked over, its pieces going into cheaper and more efficient technologies down the road? The answer may depend in part on IBM's rivals.

Google is the natural competitor, but it comes at the problem from an entirely different angle. While IBM builds a machine to grapple with one question at a time, Google is serving much of networked humanity. Electricity is a major expense, and even a small increase of wattage per query could put a dent in its profits. This means that even as Google increasingly looks to respond to queries with concrete answers, it can devote to each one only one billionth the processing power of Watson, or less. Peter Norvig, the research director, said that Google's big investments in natural language and machine translation would lead the company toward more sophisticated question-answering for its mass market. As its search engine improves its language skills, he said, it will be able to carry out smarter hunts and make better sense of its users' queries. The danger for IBM isn't head-to-head competition from Google and other search engines. But as searching comes closer to providing answers to queries in English, a number of tech startups and consultants will be able to jury-rig competing question-answering machines, much the way James Fan built his Basement Baseline at the dawn of the *Jeopardy* project.

As Google evolves, Norvig said, it will start to replicate some of Watson's headier maneuvers, combining data from different sources. "If someone wants per capita income in a certain country, or in a list of countries, we might bring two tables together," he said. For that, though, the company might require more detailed queries. "It might get to the point where we ask users to elaborate, and to write entire sentences," he said. In effect, the computer will be demanding something closer to a *Jeopardy* clue — albeit with fewer puns and riddles.

This would represent a turnaround. For more than a decade, the world's Web surfers have learned to hone their queries. In effect, they've used their human smarts to reverse-engineer Google's algorithms — and to understand how a search engine "thinks." Each word summons a universe of connections. Looking at each one like a circle in a Venn diagram, the goal is to organize three or four words — 3.5 is the global average — whose circles have the smallest possible overlap. For many, this analysis has become almost reflexive. Yet as the computer gets smarter, these sophisticated users stand to get poorer results than those who type long sentences, even paragraphs, and treat the computer as if it were human.

And many of the computer systems showing up in our lives will have a far more human touch than Watson. In fact, some of the most brilliant minds in AI are focusing on engineering systems whose very purpose is to leech intelligence from people. Luis Von Ahn, a professor at Carnegie Mellon, is perhaps the world's leader in this field. As he explains it, "For the first time in history, we can get one hundred or two hundred million people all working on a project together. If we can use their brains for even ten or fifteen seconds, we can create lots of value." To this end, he has dreamed up online games to attract what he calls brain cycles. In one of them, the ESP game, two Web surfers who don't know each other are shown an image. If they type in the same word to describe it, another image pops up. They race ahead, trying to match descriptions and finish fifteen images in two and a half minutes. While they play, they're tagging photographs with metadata, a job that computers have not yet mastered. This dab of human intelligence enables search engines to find images. Von Ahn licensed the technology to Google in 2006. Another of his innovations, ReCaptcha, presents squiggly words to readers, who fill them in to enter Web sites or complete online purchases. By typing the distorted letters, they prove they're human (and not spam engines). This is where the genius comes in. The ReCaptchas are drawn from the old books in libraries. By completing them, the humans are helping, word by crooked word, to digitize world literature, making it accessible to computers (and to Google, which bought the technology in 2009).

This type of blend is likely to become the rule as smarter computers spread into the marketplace. It makes sense. A computer like Watson, after all, is an exotic beast, one developed at great cost to play humans in a game. The segregated scene on the *Jeopardy* stage, the machine separated from the two men, is in fact a contrivance. The question-answering contraptions that march into the economy, Watson's offspring and competitors alike, will be operating under an entirely different rubric: What works and at what cost? The winners, whether they're hunting for diseases or puzzling out marketing campaigns, will master different blends. They'll figure out how to turbocharge thinking machines with a touch of human smarts and, at the same time, to augment human reasoning with the speed and range of machines. Each side has towering strengths and glaring vulnerabilities. That's what gives the *Jeopardy* match its appeal. But outside the *Jeopardy* studio, stand-alones make little sense.

10

How to Play the Game

THE TIME FOR BIG fixes was over. As the forest down the hill from the Yorktown lab took on its first dabs of yellow and red, researchers were putting the finishing touches on the question-answering machine. On the morning of September 10, 2010, five champion *Jeopardy* players walked into the Yorktown labs to take on a revamped and invigorated Watson. IBM's PR agency, Ogilvy, had a film crew in the studio to interview David Ferrucci and his team during the matches. The publicists were not to forget that focus of the campaign, which would extend into television commercials and Web videos over the coming months, would be on the people behind the machine. Big Blue was about people. That was the message. And the microphones on this late summer day would attempt to capture every word.

Over the previous four months, since the end of the first round of sparring sessions, Watson's creators had put their machine through a computer version of a graduate seminar. Watson boasted new algorithms to help sidestep disastrous categories—so-called train wrecks. Exhaustive new fact-checking procedures were in place to guide it to better responses in Final Jeopardy, and it had a profanity filter to steer it away from embarrassing gaffes. Also, it now received the digital read of *Jeopardy* answers after each clue so it could learn on the fly. This new intelligence clued Watson into its rivals' answers. It was as if the deaf machine had sprouted ears. It also sported its new finger. Encased in plastic, the apparatus gripped a *Jeopardy* buzzer and plunged it with its metal stub in three staccato bursts when Watson had enough confidence to bet. Even Watson's body was new. Over the summer, Eddie Epstein and his team had moved the entire system to IBM's latest generation of Power 7 Servers. If Watson was going to promote the company, it had to be running on the hardware Big Blue was selling.

In the remaining months leading up to the match against Ken Jennings and Brad Rutter, most of the adjustments would address Watson's game strategy: which categories to pick and how much to wager. It was getting too late to lift the machine's IQ. If Watson misunderstood clues and botched answers, they'd have to live with it. But the researchers could continue to fine-tune its betting strategy. Even at this late date, Watson could learn to make smarter decisions.

Though the final match was only months away, the arrangements between *Jeopardy* and IBM remained maddeningly fluid. An agreement was in place, but the contract had not yet been signed. Rumors about the match spread wildly on the Quiz Bowl circuits, yet the command from Culver City was to maintain secrecy. Under no circumstances were the names of the two participants to be released, not even the date of the match. On his blog, Jennings continued with his usual word games, stories about his children, and details of a trip to Manchester, England, which sparked connections in his fact-swimming mind to songs by Melissa Manchester and one from the musical *Hair* ("Manchester, England, across the Atlantic Sea . . ."). Nothing about his upcoming encounter with Watson.

Behind the scenes, *Jeopardy* officials maneuvered to get Jennings and Rutter a preview of this digital foe they'd soon be facing. Could they visit the Yorktown labs to see Watson in action, perhaps in early November? This inquiry led to further concerns. If the humans saw Watson and its weaknesses, they'd know what to prepare for. Ferrucci worried that they would focus on its electronic answer panel, which showed its top five responses to every clue. "That's a look inside its brain," he said. One Friday, as a sparring match took place in the *Jeopardy* lab and visiting computer scientists from universities around the country cheered Watson on, Ferrucci stood to one side with Rocky Schmidt and discussed just how much Jennings and Rutter would see—if they were granted access at all.

It was during this period that surprising news emerged from *Jeopardy*. A thirty-three-year-old computer scientist from the University of Delaware, Roger Craig, had just broken Ken Jennings's one-game scoring record with a $77,000 payday. "This Roger Craig guy," Jennings blogged a day later, from England, "is a monster. . . . I only wish I could have been in the *Jeopardy* studio audience to cheer him on in person, like Roger Maris's widow or something. Great great stuff." Jennings, like Craig himself, noted that Craig shared the name of a San Francisco 49er running back from the great Super Bowl squads of the 1980s. (*Jeopardy* luminaries recite such facts as naturally as the rest of us breathe or sweat. They can hardly help themselves.) Craig went on to win $231,200 over the course of six victories. What distinguished him more than his winnings were his methods. As a computer scientist, he used the tools of his trade to prepare for *Jeopardy*. He programmed himself, optimizing his own brain for the game. As the Watson team and the two human champions marched toward the matchup, each side busy devising its own strategy, Roger Craig stood at the intersection of the two domains.

Several weeks later, in mid-October, Craig sat at a pub in Newark, Delaware, discussing his methods over multiple refills of iced tea. With his broad face, wire-rimmed glasses, and a hairline in retreat, Craig looked the part of a cognitive warrior. Like many *Jeopardy* champions, he had spent his high school and college years in Quiz Bowl competitions and stuck with it even for the first couple of years of his graduate schooling at the University of Delaware. He originally studied biology, with the idea of becoming a doctor. But like Ferrucci, he had veered from medicine into computing. "I realized I didn't like the sight of blood," he said. After a short stint researching plant genomics at Dupont, he went on to study computational biology at the computer science school at Delaware. When he appeared on *Jeopardy*, he was within months of finishing his dissertation, which featured models of protein interactions within a cell. This, he hoped, would soon land him a lofty research post in a pharmaceutical lab or academia. But it also provided him with the know-how and the software tools for his hobby, and he easily created software to train himself for *Jeopardy*. "It's nice to know how to program. You get some Perl scripts," he said, referring to a popular programming language. "Then it's just chop, chop, chop, boom!"

Much like the researchers at IBM, Craig divided his personal *Jeopardy* program into steps. First, he said, he developed the statistical landscape of the game. Using sites like J! Archive, he could calculate the probability that certain categories, from European capitals to anagrams, would pop up.

Mapping the *Jeopardy* canon, as he saw it, was simply a data challenge. "Data is king," he said. Then, with the exacting style of a *Jeopardy* champ, he corrected himself. "It should be data *are* king, since it's plural. Or I guess if you go to the Latin, *Datum* is king . . ."

The program he put together tested him on categories, gauged his strengths (sciences, NFL football) and weaknesses (fashion, Broadway shows), and then directed him toward the preparation most likely to pay off in his own match. To patch these holes in his knowledge, Craig used a free online tool called Anki, which provides electronic flash cards for hundreds of fields of study, from Japanese vocabulary to European monarchs. The program, in Craig's words, is based on psychological research on "the forgetting curve." It helps people find holes in their knowledge and determines how often they need those areas to be reviewed to keep them in mind. In going over world capitals, for example, the system learns quickly that a user like Craig knows London, Paris, and Rome, so it might spend more time reinforcing the capital of, say, Kazakhstan. (And what would be the Kazakh capital? "Astana," Craig said in a flash. "It used to be Almaty, but they moved it.")

At times, the results of Craig's studies were uncanny. His program, for example, had directed him to polish up on monarchs. One day, looking over a list of Danish kings, he noticed that certain names repeated through the centuries. "I said, 'OK, file that away,'" he recalled. (Psychologists call such decisions to tag certain bits of information for storage "judgments of learning." *Jeopardy* players spend many of their waking hours forming such judgments.) In his third *Jeopardy* game, aired on September 15, Craig found himself in a tight battle with Kevin Knudson, a math professor from the University of Florida. Going into Final Jeopardy, Craig led, $13,800 to $12,200. The final category was Monarchs, and Craig wagered most of his money, $10,601. Then he saw the clue: "From 1513 to 1972, only men named Christian & Frederick alternated as rulers of this nation." It was precisely the factoid he had filed away, and he was the only one who knew it was Denmark. Only days before these games were taped, in mid-July, Craig had seen the sci-fi movie *Inception,* in which Leonardo DiCaprio plunges into dream worlds. "I really wondered if I was dreaming," he said. After three matches, it was lunchtime. Roger Craig had already pocketed $138,401.

Craig had been following IBM's *Jeopardy* project and was especially curious about Watson's statistically derived game strategy. He understood that language processing was a far greater challenge for the IBM team. But as a human, Craig had language down. What he didn't have was a team of Ph.D.s to run millions of game simulations on a cluster of powerful computers. This would presumably lead to the ideal strategy for betting and picking clues at each step of the game. His interest in this was hardly idle. By winning his six games, Craig would likely qualify for *Jeopardy*'s Tournament of Champions in 2011. Watson's techniques could prove invaluable. As soon as his shows had aired in mid-September (and he was free to discuss his victories), he e-mailed Ferrucci, asking for a chance to IBM and spar with Watson. Ferrucci's response, while cordial, was noncommittal. *Jeopardy*, not IBM, was in charge of selecting Watson's sparring partners.

Before going on *Jeopardy*, Craig had long relied on traditional strategies. He'd read books on the game, including the 1998 *How to Get on Jeopardy—And Win,* by Michael DuPee. He'd also gone to Google Scholar, the search engine's repository of academic works, and downloaded papers on Final

Jeopardy betting. Craig was steeped in the history and lore of the games, as well as various strategies, many of them named for players who had made them famous. One Final Jeopardy technique, Marktiple Choice, involves writing down a number of conceivable answers and then eliminating the unlikely ones. Formulated by a 2003 champion, Mark Dawson, it prods players to extend the search beyond the first response that pops into their mind. (In that sense, it's similar to the more systematic approach used by Watson.) Then there's the Forrest Bounce, a tactic named for a 1986 champion, Chuck Forrest, who disoriented his foes by jumping from one category to the next. "You can confuse your opponents," said Craig, who went on to use the technique. (This irked even some viewers. On a *Jeopardy* online bulletin board, one North Carolinian wrote, "I could have done without Roger winning . . . I can't stand players that hop all over the board. It drives me nuts.")

When it came to *Jeopardy*'s betting models, Craig knew them cold. One standard in the Final Jeopardy repertoire is the two-thirds rule. It establishes that a second-place player with at least two-thirds the leader's score often has a better chance to win by betting that the leader will botch the final clue (which players do almost half the time). Say the leader going into Final Jeopardy has $15,000 and the second-place player has $10,000. To ensure a tie for victory (which counts as a win for both players), the leader must bet at least $5,000. Otherwise, the number two could bet everything, reach $20,000, and win. But missing the clue, and losing that $5,000, will drop the leader into a shared victory with the second-place player—if that player bets nothing. This strategy often makes sense, Craig said, because of the statistical correlation among players. He hadn't run nearly as many numbers as the IBM team, but he knew that if one player missed a Final Jeopardy clue, it was probably a hard one, and the chances were much higher that others would miss it as well.

Craig bolstered his *Jeopardy* studies with readings on evolutionary psychology and behavioral economics, including books by Dan Ariely and Daniel Kahneman. They reinforced what he already knew as a poker player: When it comes to betting, most people are scared of losing and bet too small. (In *Jeopardy*'s lingo, which some might consider sexist, timid bets are "venusian," audacious ones, "martian.")

Craig would tilt strongly toward Mars. In his first game, he held a slender lead when he landed on a Daily Double in the category Elemental Clues. The previous clues in the group all featured symbols for elements in the periodic table. Craig didn't know all hundred and eight of them, but as a scientist he was confident that he'd know any that would be featured on *Jeopardy*. He said he was "95 percent sure" that he'd come up with the right answer, so he bet all of his money, $12,400. It turned out to be the largest bet since one placed by Ken Jennings six years earlier. The clue was "PD. A great place to hear music." For the scientist, it was a cinch. "Palladium," Craig said, recalling his golden moment. "Boom. Twenty-four thousand dollars."

That was when he made what he called his rookie mistake, one he was convinced Watson would avoid. His palladium clue was the first Daily Double of the two in Double Jeopardy. Another one lurked somewhere on the board, and he forgot about it. For the leader in *Jeopardy*, Daily Doubles represent danger, for they can lift a trailing player back into contention. So a leader who controls the board, as he did, should hunt down the remaining Daily Double. They tend to be in higher-dollar rows, where the clues are more difficult. Craig seemed to be on the verge of winning in a romp. With only seconds left in the round, he led his closest competitor, a medievalist from Los Angeles named Scott Wells, by a commanding $33,600 to $11,800. But he lost control of the board with a $400 clue: "On May 9, 1921, this 'letter-perfect' airline opened its first passenger office in Amsterdam." Wells beat him to the buzzer and correctly answered "What is KLM?" Then, as time ran out, he proceeded to land on the second Daily Double. Craig was mortified. "I thought I'd die," he said. Wells bet $10,000, which would put him well within striking distance in Final Jeopardy. The clue: "In 1939 this Russian took the 1st flight of a practical, single-rotor helicopter, & why not? He built the thing!" Craig survived his blunder when Wells failed to come up with "Who is Igor Sikorsky?"

As he left the Culver City studios after his first day on *Jeopardy*, Craig was experiencing a host of human sensations. First, he was euphoric. He had amassed $197,801, a five-game record. As he headed out for a bite with the fellow players he had befriended, he felt a little embarrassed. Here he was, swimming in money, and thanks to him, every one of them had crashed and burned on their once-in-a-lifetime chance to win at *Jeopardy*. Between breakfast and dinner, he had doused the dreams of ten players. Many of them had prepared for years, even decades, watching the show religiously, reading almanacs, studying flash cards, wowing friends and relatives, and envisioning that they'd be the next Ken Jennings—or at the very least stick around for a few games. Now they were heading home with a loser's pay of $1,000 or $2,000, barely enough for the plane ticket. Craig, on the other hand, might turn out to be the next superstar. It was at least a possibility. Ken Jennings had never won as much in a match or a single (five-match) day. No one had. That night, in his room at the

Radisson Hotel in Culver City (which offered limo service to the Sony lot), he tossed and turned. The next morning, while a *Jeopardy* staffer was applying makeup to the new champion's face, Craig found himself yawning. This was worrisome. The night before his magical five-game run, he recalled, he had slept soundly for nine hours. Now, he didn't feel nearly as good.

Still, Craig blitzed though his first game. His crucial clue was another jumbo bet—$12,000, this time—on a Daily Double, in which he identified "small masses of lymphoid tissue in the nasopharynx" ("What are adenoids?"). He chalked up another $34,399 and appeared to be off and running.

But the next match was his undoing. He faced Matt Martin, a police officer from Arlington, Virginia, and Jelisa Castrodale, a sportswriter from North Carolina. Just a day earlier, his luck with Danish kings and atomic elements made him wonder if he was dreaming. Now his fortunes took a cruel turn. Sleep-deprived, he found himself struggling in a category that seemed to be mocking him: "Pillow talk." Such fluff was hardly his forte. Castrodale identified the "small scattered pillows also known as scatter cushions" ("What is a throw pillow?") and the "child carrying the pillow in a wedding procession" ("What is a ring-bearer?"). And when a clue asked about "folks with tailbone injuries" sitting on "pillows in the shape of these sweet baked treats," Martin buzzed in. It was the cop, as Alex Trebek gleefully noted, who answered, "What are donuts?"

Barely a week before Craig's final show aired, Watson was engaged in a closely fought match with the former champion Justin Bernbach, and they were playing the very same clues. This was the day that Watson, following a dominating morning, later faltered and crashed. Its patterns in this game seemed to mirror those of Roger Craig. Like Craig, Watson appeared largely lost on pillow talk. Both of them, however, swept through the category on the ancient civilization of Ur. (When you have a category like that, Craig later explained, "You almost know the answers before they ask the questions." He listed a few on the fingers of one hand: Iraq, Sumeria, Cyrus the Great, and Ziggurats (the terraced monuments they built). "What else can they ask about Ur?" Watson, though following a different logic, delivered the same winning results. Watson and Craig also thrived in the category "But what am I?" It featured the Latin names for certain animals, along with helpful hints, alerting players, for example, not to confuse "cyanocitta cristata" with Canadian baseball players ("What are Blue Jays?"). These were easy factoids for computer and computer scientist alike.

As Watson went into Final Jeopardy on that September afternoon, it held a slim lead over Bernbach and a comfortable one over Maxine Levaren, a personal success coach from San Diego. But it lost the game to Bernbach, you might recall, by missing a clue in the category Sports and the Media. It failed to name the city whose newspaper celebrated the previous February 8 with the headline: "Amen! After 43 Years, Our Prayers Are Answered." The computer had only 13 percent confidence in Chicago, but that was higher than its confidence in its other candidates, including Omaha and two cities associated with prayer, Jerusalem and the Vatican. In retrospect, Watson was scouring its database for events dated February 8. But the machine, raised in the era of instant digital news, ignored the lag at the heart of traditional paper headlines: Most of the events they describe occurred the previous day.

Like Watson, Roger Craig reached Final Jeopardy cling-

ing to a narrow lead, $22,000 to $19,700, over Jelisa Castrodale. The Sports and the Media category looked perfect for
the sportswriter. But Craig was a fan as well and a master of
sports facts—especially those concerning football. The same
clue Watson had botched, featuring forty-three years and answered prayers, popped up on the board and the contestants
wrote their responses. After the jingle, Alex Trebek turned to
them. Martin, who lagged far behind, incorrectly guessed:
"What is Miami?" Castrodale was next: "What is New Or-
leans?" That was right. She had bet all but one dollar, which
lifted her to $39,339. Craig had anticipated her bet and topped
it: He would win by $2 if he got it right. But his answer was
840 miles off target. The six-time champion, who had trained
himself with the methods and rigor of computer science,
came up with the same incorrect response as his electronic
role model: "What is Chicago?"

Was the melding of man and machine leading Craig and
Watson through the same thought processes and even to the
same errors? Weeks later, sitting in IBM's empty *Jeopardy* studio, David Gondek opened his Mac and traced the cognitive route that led Watson to the Windy City. "It really didn't
have any idea," he said, clicking away. The critical document,
Gondek found, turned out to be news about a prayer meeting in Chicago on February 8, which featured a prominent
swami. When Watson failed to come up with convincing responses, which correlated, statistically and semantically, to the
clue, it turned to documents like this one with a few matching words. The machine had negligible confidence in answers
from such sources. But in this case, the machine had no better
option.

Craig had a different story. In the thirty seconds he had to
mull Final Jeopardy, thoughts about a prayer service featuring

a swami in Chicago never entered his mind. But his analysis, usually so disciplined, was derailed by an all-too-human foible. He fell to suggestion, one nourished by his environment. Just a short drive north of his home in Delaware, the ice hockey team in Philadelphia, the Flyers, had recently battled to the finals of the Stanley Cup. This awakened hockey fever in the metropolitan area and an onslaught of media coverage, along with endless chatter and speculation. Hockey hadn't been on people's minds to this degree since the glory years of the franchise, when the "Broad Street Bullies" won back-to-back cups in the mid-1970s. The Flyers ultimately lost to the Chicago Black Hawks, a team that hadn't won in forty-nine years (six years longer than the Saints). So even though Craig was a "huge football fan" who hadn't missed watching a Super Bowl since his childhood, he had hockey in his head when he saw the Final Jeopardy clue. Much like the psychology test subjects who mistook Moses for the animal keeper on the ark, Craig focused on a forty-something-year championship drought—and looked right past the crucial February date. The hockey final, after all, had been in June. "I blew it," he said. So did Watson. But despite their virtuoso talents and similar techniques, in this one example of failure they each remained true to their kind. One was dumb as only a machine can be, the other human to a fault.

During the sparring sessions in the spring, Watson had relied on simple heuristics to guide its strategy. Ferrucci at one point called it brain dead, and David Gondek, who had written the rules, had to agree. You might say that such heuristics are "brain-dead by definition," he said, since they replace analysis with rules. But what a waste it was to equip Watson, a machine that could carry out billions of calculations per second, with such a rudimentary set of instructions.

There was no reason, of course, for Watson's strategy to be guided by a handful of simple rules. The machine had plenty of processing power, enough to run a trillion-dollar trading portfolio or to manage all of the air traffic in North America or even the world. Figuring out bets for a single game of *Jeopardy* was well within its range. But before the machine could become a strategic whiz, Gondek and his team had to turn thousands of *Jeopardy* games into a crazy quilt of statistical probabilities. Then they had to teach Watson—or help it teach itself—how best to play the game. This took time.

The goal was to have Watson analyze a dizzying assortment of variables, from its track record on anagrams or geography puzzlers to its opponents' ever-changing scores. Then it would come up with the ideal betting strategy for each point of the game and for each clue. This promised to be much simpler for Watson than the rest of its work. English, after all, was foreign to the machine, and *Jeopardy* clues, even after years of work, remained challenging. Game strategy, with its statistical crunching of probabilities, played to Watson's strengths.

To tutor Watson in the art of strategy, Gondek brought in one of IBM's gaming masters, an intense computer scientist named Gerald Tesauro. Short, dark, and neatly dressed, his polo shirt tucked cleanly into dark slacks, Tesauro was one of the more competitive members of the *Jeopardy* team. He took pride, for example, in his ability to beat Watson to the buzzer. Once, in a practice match against the machine, he managed to buzz in twenty-four times, he later said, and got eighteen of the clues right. Like a basketball player who's hitting every shot, he said, he was "in some kind of a zone" (though, to be honest, that 75 percent precision rate would place him in a crowd of *Jeopardy* also-rans). Even when Tesauro was in the audience, he would play along in his mind, jerking an imaginary buzzer in his fist each time he knew the response.

Tesauro gained global renown in the '90s when he developed the computer that mastered the five-thousand-year-old game of backgammon. (Sumerians, as Roger Craig may already know, played a variation of it in the ancient city of Ur.) What distinguished Tesauro's approach was that he didn't teach the machine a thing. Using neural networks, his system, known as TD-Gammon, learned on its own. Following Tesauro's instructions, it played games against itself, millions of them. Each time it won or lost, it drew conclusions. Certain moves in certain situations led more often to victory, others to defeat. Although this was primitive feedback—no more than thumbs up, thumbs down—each game delivered a minuscule improvement, Tesauro said. Over the course of millions of games, the machine developed a repertoire of winning moves for countless scenarios. Tesauro's machine beat champions.

Tesauro's first goal was to run millions of simulated *Jeopardy* games, just as he had with backgammon. For this he needed mathematical models of three players, Watson and two humans. Modeling Watson wasn't so hard. "We knew all of its algorithms," he said, and the team had precise statistics on every aspect of its behavior. The human players were more complicated. Tesauro had to pull together statistics on the thousands of humans who had played *Jeopardy*: how often they buzzed in, their precision in different levels of clues, their betting patterns for Daily Doubles and Final Jeopardy. From these, the IBM team pieced together statistical models of two humans.

Then they put them into action against the model of Watson. The games had none of the life or drama of *Jeopardy*—no suspense, no jokes, no jingle while the digital players came up with their Final Jeopardy responses. They were only simulations of the scoring dynamics of *Jeopardy*. Yet they were valuable. After millions of games, Tesauro was able to calculate the value of each clue at each state of the game. If Watson was in second place, trailing by $1,500 with $14,400 left on the board, what size bet on a Daily Double maximized its chance of winning? The answer changed with every move, and Tesauro was mapping it all out. Humans, when it came to betting, only had about five seconds to make a decision. They went with their gut. Watson, like its number-crunching brethren in advertising and medicine, was turning its pile of data into science.

The science, it turned out, was a bit scary. Watson's model was based on the record it had established following the simple heuristics. And studies showed that the machine, much like risk-averse humans, had been dramatically underbetting. In many stages of the game, according to Tesauro's results, the computer could maximize its chances by wagering nearly everything it had. (This wasn't always the case. If Watson enjoyed a big lead late in the game, it made sense to minimize a bet.) When Tesauro adjusted Watson's strategy toward a riskier blend of bets, it started winning more of the simulated games. He and Gondek concluded that in many scenarios, Watson should bet the farm. "When we first went to Dave Ferrucci about this," Tesauro recalled, "he turned pale as a sheet and said, 'You want to do what?'"

"We showed him all the extra wins we were getting with this," Gondek said. "But he looked at the colossal bets we were making and said, 'What if you get them wrong?'"

The conflict between rational analysis and intuition were

playing out right in the IBM War Room. And Ferrucci, much like the humans who placed small, safe bets every evening on *Jeopardy*, was turning away from statistics and focusing on a deeper and more primitive concern: survival. Watson was going to be playing only one game on national television. What if it bet big on that day and lost?

"That would really look bad for us," Tesauro said. Perhaps it would be better to sacrifice a win or two out of a hundred and protect Watson a bit more from prime-time catastrophe. It wasn't clear. The strategy team continued to crunch numbers.

The numbers flowing in from the real matches, where Watson was playing flesh-and-blood humans, were improving. Through the autumn season, the newer, smarter Watson powered its way past scores of *Jeopardy* champions. It won nearly 70 percent of its matches; its betting was bolder, its responses more assured. It still crashed from time to time, of course, and routinely made crazy mistakes. On one Daily Double, it was asked to name the company that in 2002 "came out with a product line featuring 2-line Maya Angelou poems." Watson missed the answer ("What is Hallmark?") and appeared to pay tribute to its creators, responding: "What is IBM?"

Watson's greatest weakness was in Final Jeopardy. According to the statistics, after the first sixty clues, Watson was leading an astounding 91 percent on the games. Yet that final clue, with its more difficult wording and complex wagering dynamics, lowered its winning percentage to 67 percent. Final Jeopardy turned Watson from a winner to a loser in one-quarter of the games. This was its vulnerability going into the match, and it would no doubt rise against the likes of Ken Jennings and Brad Rutter. The average human got Final Jeopardy right about half the time, according to Gondek. Watson hovered just below 50 percent. Ken Jennings, by contrast, aced Final Jeopardy clues at a *68 percent* rate. That didn't bode well for the machine.

Brad Rutter, undefeated in his *Jeopardy* career, walked into the cavernous *Wheel of Fortune* studio. It was mid-November, just two months before he and Ken Jennings would take on Watson. Rutter, thirty-two, is thin and energetic, with sharply chiseled features. His close-cut black beard gives him the look of a vacationing television star. This is appropriate, since he recently moved from his native Lancaster, Pennsylvania, to L.A.'s Beechwood Canyon, right under the famous Hollywood sign. He's trying to make it as an actor.

On this autumn day, Rutter and Jennings were having their orientation for the upcoming match. They were shuttling back and forth between meetings in the Robert Young Building and interviews in the empty *Wheel of Fortune* studio. Rutter, clearly fascinated by television, spotted a rack of men's suits by the stage. "Are those Pat Sajak's?" he asked, referring to the longtime *Wheel of Fortune* host. Told that they were, he went over to check the labels. For years, the show announced every evening that Sajak's wardrobe was provided by Perry Ellis. Rutter, a stickler for facts, wanted to make sure it was true. It was.

The previous evening, Rutter had been given a Blu-ray Disc featuring five of Watson's sparring rounds. He studied them closely. He noticed right away that Watson hopped around the board, apparently hunting for Daily Doubles. He also focused on Watson's buzzer speed and was relieved to see that humans often managed to beat the machine. This was crucial for Rutter, who viewed speed as his greatest advantage.

He said he was no expert on computers and had only a vague idea of how Watson worked. But he had expected the IBM team to give Watson an intricate timing program to anticipate the buzz. (This was a frightfully complex option that Ferrucci had decided not to pursue.) "That scared me," Rutter said.

Rutter's speed is legendary. It fueled his 16-0 record on *Jeopardy*, including his decisive victories over Jennings. It was such an advantage that IBM's Gondek referred to Rutter as "Jennings Kryptonite." Rutter said he wasn't sure what made his thumb so fast, but he had a theory. "I used to play a lot on the Nintendo entertainment system when I was a kid," he said. "And if you played Super Mario Brothers or Metroid, you had to hit the button to jump at exactly the right time. It was not about speed but timing. And that's what the *Jeopardy* buzzer is about." This meant that a computer game trained the human, who would later use those same skills to take on another computer.

But Rutter boasted strengths beyond mere speed. In an Ultimate Tournament of Champions match that aired in May 2005, he found himself in a most unusual position—third place—heading into Final Jeopardy. The category was People and Places, and the clue: "This Mediterranean island shares a name with President Garfield's nickname for his wife."

"I started scanning Mediterranean islands," Rutter said. "OK. Sardinia? No. Corsica? No. Sicily? No. Menorca? Mallorca? Malta?" He figured, "Malta could be a girl's name," and wrote it down. But he knew it was wrong. As the theme music played, he continued to think of islands. Lesbos, Rhodes, Ibiza . . . "With about five seconds left," he said, "I got to Crete." All at once the pieces came together. "Crete could be short for Lucretia. That's a very nineteenth-century name. And then it was an apparition in my head. I'd looked at a list of First Ladies, and somehow Lucretia Garfield popped out at me. I can't explain it. So I scribbled down Crete. It was barely legible. I was the only one to get it right, and I ended up winning by a dollar."

A timely spark of human brilliance had saved him. It featured a series of insights and connections Watson would be hard-pressed to match. Indeed, in the coming showdown, Rutter's competition on that type of clue was more likely to come from the other human on the stage. This led to a question: Was it fair that the two humans had to battle each other in addition to the machine? Mightn't it be easier for one player to face two machines?

Rutter thought so. "I've seen Ken play seventy-four matches," he said. "I know his strengths and weaknesses pretty well. They're different than Watson's. So when I'm picking different categories or clues off the board, who do I attack? Whose weaknesses do I try to get to? That's a tough question. I haven't really figured it out yet. I'm going to be thinking about it a lot."

Jennings, sitting in the same studio, elaborated on the point. "I don't mean it to sound like I'm making excuses already," he said, "but there is some inherent disadvantage that there are two humans and one Watson." The way he saw it, Watson's algorithms would master "a certain percentage of the *Jeopardy* canon." And if the computer was fast on the buzzer, it would dominate in those areas. That left two players to battle over the clues "that only humans can do."

The thirty-six-year-old Jennings, with his featherweight build, is far smaller than Rutter. He has an easy laugh and a self-effacing style. He hadn't yet found a Blu-ray player, he said, to watch the video of Watson in action. But he had clearly been reading everything he could find about the computer, including technical articles. Jennings double-majored in computer science and English at Brigham Young University and later worked as a computer programmer in Salt Lake City. When he heard about the match against Watson, he said, it excited him. "I said, 'Wow, we get to see if a computer can play *Jeopardy,*'" he said. "I was more interested in the geeky part."

As Jennings studied up on Watson's algorithms and its massive parallel processing, he couldn't help comparing the computer to his own mind. "Many of the tricks that I used in *Jeopardy* are things that I read Watson does," he said.

He gave an example. One *Jeopardy* clue asked for the name of two of Jesus' disciples whose names are both top-ten baby male names and end in the same letter. "I remember thinking," he said, "that that's not the kind of thing you can know. The only way to do it is to break it down, make a list, do the Venn diagram of it. And I come to find out that Watson does exactly that. It's very good at decomposing questions, so it does the two fact sets in parallel. Then it does the Venn diagram to see if there's anything on both lists." Jennings paused for a moment, then said, "Matthew and Andrew, by the way. I got that one right at the last minute. I was about to put James and Judas, but I don't think Judas is a popular baby name, for some reason . . ."

Jennings reflected on traveling across the country, to IBM's lab in Yorktown, to take on Rutter and Watson. "It's a little different to be the road team," he said. "I'm not playing in the familiar studio where I have the muscle memory and the good times and the million-dollar check. I'm picturing a very sterile lab from the fifties, people running around in white coats. . . ."

Jennings had been following Watson's record against its

sparring partners, and the trend looked worrisome. In the beginning of the matches, he said, Watson was winning 64 percent of the time against standard *Jeopardy* players. "Now they've fine-tuned it, and it's 67 percent against Tournament of Champions players. I know it can still be beat," he said. "But I think to myself: Could I win 67 percent of my games against Tournament of Champions players? That's not something I've ever done. I rattled off a very long streak, but it was against rookie players." He said that he was going into the match feeling, for the first time, like an underdog.

Jennings is well known for his disarming modesty. In previous games, it could be argued, it may have benefited him as a psychological tactic. Rivals encountered a likable and unassuming young man who seemed almost surprised at his own success. By the time they looked at the scoreboard, he was annihilating them.

Such tactics wouldn't mean much in the coming showdown. Jennings and Rutter would be facing a foe impervious to nerves and psychological maneuvering. And while millions tuned in to what promised to be an epic knowledge battle between two men and a thinking machine, the drama would leave Watson unmoved. The machine, unlike everyone else, had no stake in the outcome.

11

The Match

DAVID FERRUCCI HAD driven the same stretch hundreds of times from his suburban home to IBM's Yorktown labs, or a bit farther to Hawthorne. For fifteen or twenty minutes along the Taconic Parkway he went over his endless to-do list. How could his team boost Watson's fact-checking in Final Jeopardy? Could any fix ensure that the machine's bizarre speech defect would never return? Was the pun-detection algorithm performing up to par? There were always more details, plenty to fuel both perfectionism and paranoia—and Ferrucci had a healthy measure of both.

But this January morning was different. As he drove past frozen fields and forests, the pine trees heavy with fresh snow, all of his lists were history. After four years, his team's work was over. Within hours, Watson alone would be facing Ken Jennings and Brad Rutter, with Ferrucci and the machine's other human trainers reduced to spectators. Ferrucci felt his eyes well up. "My whole team would be judged by this one game," he said later. "That's what killed me."

The day before, at a jam-packed press conference, IBM had unveiled Watson to the world. The event took place on a glittering new *Jeopardy* set mounted over the previous two weeks by an army of nearly a hundred workers. It resembled the set in Culver City: the same jumbo game board to the left, the contestants' lecterns to the right, with Alex Trebek's podium in the middle. In front was a long table for the *Jeopardy* officials, where Harry Friedman would sit, Rocky Schmidt at his side, followed by a line of writers and judges, all of them with monitors, phones, and a pile of old-fashioned reference books. Every piece was in place. But this East Coast version was plastered with IBM branding. The shimmering blue wall bore the company's historic slogan, Think, in a number of languages. Stretched across the shiny black floor was a logo that looked at first like Batman's emblem. But closer study revealed the planet Earth, with each of the continents bulging, as if painted by Fernando Botero. This was Chubby Planet, the symbol of IBM's Smarter Planet campaign and the model for Watson's avatar. In the negotiations with *Jeopardy* over the past two years, IBM had lost out time and again on promotional guarantees. It had seemed that Harry Friedman and his team held all the cards. But now that the match was assured and on Big Blue's home turf, not a single branding opportunity would be squandered.

The highlight of the press event came when Jennings and Rutter strode across the stage for a five-minute, fifteen-clue demonstration. In this test run, Watson held its own. In fact, it ended the session ahead of Jennings, $4,400 to $3,400. Rutter trailed with $1,200. Within hours, Internet headlines proclaimed that Watson had vanquished the humans. It was as if the game had already been won.

If only it were true. The demo match featured just a handful of clues and included no Final Jeopardy—Watson's Achilles' heel. What's more, after the press departed that afternoon,

Watson and the human champs went on to finish that game and play another round—"loosening their thumbs," in the language of *Jeopardy*. In these games Ferrucci saw a potential problem: Ken Jennings. It was clear, he said, that Jennings had prepped heavily for the match. He had a sense of Watson's vulnerabilities and an aggressive betting strategy specially honed for the machine. Brad Rutter was another matter altogether. Starting out, Ferrucci's team had been more concerned about Rutter than Jennings. His speed on the buzzer was the stuff of legend. Yet he appeared relaxed, almost too relaxed, as if he could barely be bothered to buzz. Was he saving his best for the match?

In the first of the two practice games, Jennings landed on all three Daily Doubles. Each time he bet nearly everything he had. This was the same strategy Greg Lindsay had followed to great effect in three sparring games a year earlier. The rationale was simple. Even with its mechanical finger slowing it down by a few milliseconds, Watson was lightning fast on the buzzer. The machine was likely to win more than its share of the regular *Jeopardy* clues. So the best chance for the humans was to pump up their winnings on the four clues that hinged on betting, not buzzing: the Daily Doubles and Final Jeopardy. Thanks to his aggressive betting, Jennings ended the first full practice game with some $50,000, a length ahead of Watson, which scored $39,000. Jennings was fired up. When he clinched the match, he pointed to the computer and exclaimed, "Game over!" Rutter finished a distant third, with about $10,000. In the second game, Jennings and Watson were neck and neck to the end, when Watson edged ahead in Final Jeopardy. Again, Rutter coasted to third place. Ferrucci said that he and his team left the practice rounds thinking, "Ken's really good—but what's going on with Brad?"

When Ferrucci pulled in to the Yorktown labs the morning of the match, the site had been transformed. The visitors' parking lot was cordoned off for VIPs. Security guards noted every person entering the building, checking their names against a list. And in the vast lobby, usually manned by one lonely guard, IBM's luminaries and privileged guests circled around tables piled with brunch fare. Ferrucci made his way to Watson's old practice studio, now refashioned as an exhibition room. There he gave a half-hour talk about the computer to a gathering of IBM clients, including J. P. Morgan, American Express, and the pharmaceutical giant Merck & Co. Ferrucci recalled the distant days when a far stupider Watson responded to a clue about a famous French bacteriologist by saying: "What is 'How Tasty Was My Little Frenchman'?" (That was the title of a 1971 Brazilian comedy about cannibals in the Amazon.)

His next stop, the makeup room, revealed his true state of mind. The makeup artist was a woman originally from Italy, like much of Ferrucci's family. As she began to work on his face she showered him with warmth and concern—acting "motherly." This rekindled his powerful feelings about his team and the end of their journey, and before he knew it, tears were streaming down his face. The more the woman comforted him, the worse it got. Ferrucci finally stanched the flow and got the pancake on his face, but he knew he was a mess. He hunted down Scott Brooks, the lighthearted press officer. Maybe some jokes, he thought, "would take the lump out of my throat." Brooks laughed and warned him that people might compare him to the new U.S. Speaker of the House, John Boehner, whose frequent tears had recently earned him the sobriquet "Weeper of the House."

This irritated the testy Ferrucci and, to his relief, knocked him out of his fragile mood. He joined his team for a last lunch, all of them seated at a long table in the cafeteria. As they were finishing, just a few minutes before 1 P.M., a roaring engine interrupted their conversations. It was IBM's chairman, Sam Palmisano, landing in his helicopter. The hour had come. Ferrucci walked down the sunlit corridor to the auditorium.

Ken Jennings woke up that Friday morning in the Crown Plaza Hotel in White Plains. He'd slept well, much better than he usually did before big *Jeopardy* matches. He had good reason to feel confident. He had destroyed Watson in one of the practice rounds. Afterward, he said, Watson's developers told him that the game had featured a couple of "train wrecks" — categories in which Watson appeared disoriented. Children's Literature was one. For Jennings, train wrecks signaled the machine's vulnerability. With a few of them in the big match, he could stand up tall for humans and perhaps extend his legend from *Jeopardy* to the broader realm of knowledge. "Given the right board," he said, "Watson is beatable." The stakes were considerable. While IBM would give all of Watson's winnings to charity, a human winner would earn a half-million-dollar prize, with another half-million to give to the charity of his choice. Finishing in second and third place was worth $150,000 and $100,000, with equal amounts for the players' charities.

A little after eleven, a car service stopped by the hotel, picked up Jennings and his wife, Mindy, and drove them to IBM's Yorktown laboratory. Jennings carried three changes of clothes so that he could dress differently for each session, simulating three different days. As soon as he stepped out of the car, *Jeopardy* officials whisked him past the crush of people in the lobby toward the staircase. *Jeopardy* had cleared out a few offices in IBM's Human Resources Department, and Jennings was given one as a dressing room.

On short visits to the East Coast, Brad Rutter liked to sleep late in order to stay in sync with West Coast time. But the morning of the match, he found himself awake at seven, which meant he faced four and a half hours before the car came for him. Rutter was at the Ritz-Carlton in White Plains, about a half mile from Jennings. He ate breakfast, showered, and then killed time until 11:30. Unlike Jennings, Rutter had grounds for serious concern. In the practice rounds, he had been uncharacteristically slow. The computer had exquisite timing, and Jennings seemed to hold his own. Rutter, who had never lost a game of *Jeopardy*, was facing a flameout unless he could get to the buzzer faster.

Shortly after Rutter arrived at IBM, he and Jennings played one last practice round with Watson. To Rutter's delight, his buzzer thumb started to regain its old magic, and he beat both Jennings and the machine. Now, in the three practice matches, each of the players had registered a win. But Jennings and Rutter noticed something strange about Watson. Its game strategy, Jennings said, "seemed naive." Just like beginning *Jeopardy* players, Watson started with the easy, low-dollar clues and moved straight down the board. Why wasn't it hunting for Daily Doubles? In the Blu-ray Discs given to them in November, Jennings and Rutter had seen that Watson skipped around the high-dollar clues, hunting for the single Daily Double on the first *Jeopardy* board and the two in Double Jeopardy. Landing on Daily Doubles was vital. It gave a player the means to build a big lead. And once the Daily

Doubles were off the board, the leader was hard to catch. But in the practice rounds, Watson didn't appear to be following this strategy.

The two players were led to a tiny entry hall behind the auditorium. As the event began, shortly after 1 P.M., they waited. They listened as IBM introduced Watson to its customers. "You know how they call time-outs before a guy kicks a field goal?" Jennings said. "We were joking that they were doing the same thing to us. Icing us." Through the door they heard speeches by John Kelly, the chief of IBM Research, and Sam Palmisano. Harry Friedman, who decades earlier had earned $5 a joke as a writer for *Hollywood Squares,* delivered one of his own. "I've lived in Hollywood for a long time," he told the crowd, "so I know something about Artificial Intelligence." When Ferrucci was called to the stage, the crowd rose for a standing ovation. "I already cried in makeup," he said. "Let's not repeat that."

Finally, it was time for *Jeopardy,* and Jennings and Rutter were summoned to the stage. They walked down the narrow aisle of the auditorium, Jennings leading in a business suit and yellow tie, the taller, loose-gaited Rutter following him, his collar unbuttoned. They settled at their lecterns, Jennings on the far side, Rutter closer to the crowd. Between them, its circular black screen dancing with colorful jagged lines, sat Watson.

The show began with the familiar music. A fill-in for the legendary announcer, Johnny Gilbert (who hadn't made the trip from Culver City), introduced the contestants and Alex Trebek. Even then, Jennings and Rutter had to wait while an IBM video told the story of the Watson project. In a second video, Trebek asked Ferrucci about the machinery behind the bionic player—now up to 2,880 processing cores. Then

Trebek gave viewers a tutorial on Watson's answer panel. It would reveal the statistical confidence that the computer had in each of its top responses. It was a window into Watson's thinking.

Trebek, in fact, had been a late convert to the answer panel. Like the rest of the *Jeopardy* team, he was loath to stray from the show's time-honored formulas. People knew what to expect from the game: the precise movements of the cameras, the familiar music, voices, and categories. Wouldn't the intrusion of an electronic answer panel distract them and ultimately make the game less enjoyable to watch? Trebek raised that concern on a visit to IBM in November. But the prospect of televising the game without Watson's answer panel horrified Ferrucci. Millions of viewers, he believed, would simply conclude that the machine had been fed all the answers. They wouldn't appreciate what Watson went through to arrive at the correct response. So while Trebek was eating lunch that day, Ferrucci had his technicians take down the answer panel. When the afternoon sessions began, it took only one game for Trebek to ask for it to be restored. Later, he said, watching Watson without the panel's analysis was "boring as hell."

Finally, it was time to play. A hush settled over the auditorium. Ferrucci, sitting between David Gondek and Eric Brown, laced his hands tightly and made a steeple with his index fingers. He watched as Trebek, with a wave of his arm, revealed the six categories for the first round of *Jeopardy*. One was Literary Character APB. Trebek explained that APB stood for "all points bulletin." This clarification was lost on the deaf Watson, which irked Ferrucci and the IBM team. Other categories were Beatles People, Olympic Oddities, Name that Decade, Final Frontiers, and Alternate Meanings. None of them looked especially vexing for the computer.

Rutter had won the draw, so he started and chose Alternate Meanings for $200. "A four-letter word for vantage point," Trebek read, "or belief." Rutter, famous for his prowess with the buzzer, won this first clue and responded correctly: "What is view?"

He asked for the $400 clue in the same category. Trebek read: "Four-letter word for the iron fitting on the hoof of a horse, or a card-dealing box in a casino."

Watson won the buzz and uttered its first syllables for an audience of millions, answering correctly: "What is a shoe?" It pronounced the final word meekly, as if unsure of itself or perhaps embarrassed. Still, with that response, Watson had $400—positive winnings against the greatest of human players. That alone was a threshold that four years earlier had appeared daunting to many—including some in the audience.

With control of the board, Watson pursued the merciless strategy mapped out by David Gondek and his team. Departing from its passive approach in the practice rounds, it moved straight to the high-dollar boxes, hunting for the Daily Double. "Let's try Literary Character APB for eight hundred," Watson said. The zinging sound of space guns echoed through the auditorium, announcing that the machine, on its first try, had landed on the Daily Double. The two APPLAUSE signs flashed over the stage, but they were hardly needed. This was Watson's crowd.

In truth, the Daily Double in the first round of *Jeopardy* is not terribly important, especially this early in the game. The players at this stage have very little money to bet. It's in Double Jeopardy, when the end is in sight and the contestants have piled up much higher winnings, that a laggard can vault toward victory, winning $10,000 or even more with a single bet. Watson's Daily Double strategy was less about padding its own lead than keeping these dangerous wild cards from its rivals.

Though Watson had won only $400, *Jeopardy* rules allow players to bet the maximum dollar number on the board, or $1,000. This would risk dropping Watson's score into negative territory. But before the machine could place its bet, Alex Trebek stopped the game. His monitor had blacked out. Technicians scurried across the black stage as Jennings slumped at his lectern. Rutter, his feet crossed at the ankles, drummed his fingers. Between them, the computer's avatar traced its endless lines of blue and red, behaving much like its close cousin, the screen-saver. When it came to patience, Watson was in a league of its own.

"Ready to go," said a voice from the control booth. "Five, four, three, two, one." The crowd again heard the recording of Watson calling for the $800 clue, the sound of zinging space guns, and the applause. Then Trebek cut in live—an art he had perfected in his twenty-seven years on the show—asking Watson how much it wanted to bet. "One thousand, please," the computer said. Then it faced this clue: "Wanted for killing Sir Danvers Carew, appearance pale and dwarfish, seems to have a split personality."

Watson didn't hesitate. "Who is Hyde?"

"Hyde, yes," Trebek said. "Dr. Jekyll and Mr. Hyde." The crowd applauded. Jennings and Rutter politely joined in. This was the custom in *Jeopardy,* though such sportsmanship seemed a bit odd when standing next to a machine.

Watson didn't stop there. Beating Jennings and Rutter to the buzz, it answered clues about the Beatles' Jude, the swimmer Michael Phelps, the monster Grendel in *Beowulf,* the 1908 London Olympics, the boundaries of black holes (event horizons), Lady Madonna, and Maxwell's Silver Hammer. By the time Rutter jumped in on a clue about the Harry Potter books ("What is Voldemort?"), Watson had $5,200, far ahead of Rutter's $1,000. Jennings trailed with only $200.

It was time for a commercial break. Off camera, Trebek shook his head as he walked across the set toward Jennings and Rutter. "I can't help but wonder if Watson was sandbagging yesterday," he said. Was the computer, like a poker player holding a royal flush, masking its strength? Rutter didn't know, but he noticed that Watson's strategy had changed. "He wasn't jumping around the way he is today," he said.

"He's a hustler," Jennings said.

In fact, before the match technicians had switched Watson to its "championship" mode. This involved two changes. First, this exhibition match was a double game. The player with the highest cumulative score in the two games would win. This changed the players' strategy. Instead of following the safest path to win each game, if only by a single dollar, players had to pile up winnings. In addition to adjusting Watson's betting algorithms for double games, the IBM team directed the machine to hunt for Daily Doubles. The practice rounds, they said, were to test the machinery and the buzzer. The goal in the match was to win. These tweaks hadn't much affected Watson's scoring in this early round. The computer had simply chanced on comfortable clues, from the Beatles to black holes. That could change.

And it did. As the opening game progressed, Watson faltered. In the Final Frontiers category, it buzzed confidently on a Latin term for end, "a place where trains can also originate." But the machine picked the wrong Latin word: "What is finis?" Jennings got "terminus" on the rebound, and inched closer.

Then Watson fell into a couple of cognitive traps. The $1,000 clue under Olympic Oddities asked about "the anatomical oddity of U.S. gymnast George Eyser, who won a gold medal on the parallel bars in 1904." Jennings won the buzz and after a pause ventured: "What is . . . he was missing a hand?" That was incorrect. Watson buzzed on the rebound.

"What is leg?" it said.

"Yes," Trebek said. But before they moved to the next clue, a judge called a halt to the game. Eyser's "leg" wasn't the anatomical oddity. Instead, it was the fact that he was *missing* a leg. After five minutes of consultation onstage with the judges, Trebek, and IBM's David Shepler—Watson's advocate—the computer's response was ruled incorrect. "It was my boo-boo," Trebek told the audience. Then he redubbed his response to Watson: "No, I'm sorry I can't accept that. I needed you to say, "What is 'He's missing a leg'?"

Watson's mistake, though subtle, reflected its misreading of the lexical answer type (LAT) in the clue. Despite years of training from James Fan and others, in this example it failed to understand precisely what it was seeking. For a national audience initially wowed by the *Jeopardy* computer, it would serve as a reminder that the machine, for all its prodigious powers, could succumb to confusion. For Jennings and Rutter, the upshot was simpler. It chopped $2,000 from Watson's lead.

This was a misstep for Watson but hardly an embarrassment. That would come later, on a $1,000 clue asking about the decade that gave birth to Oreo cookies and the first modern crossword puzzle. Jennings won the buzz and answered, "What are the twenties?' This was wrong. The deaf Watson won the rebound and promptly repeated the same wrong answer. The machine, for all its brilliance, was in many aspects oblivious. This was no secret in IBM's War Room, but now the whole world could see it.

As this first *Jeopardy* round came to a close, Rutter climbed and Watson tumbled. They ended in a tie, the co-leaders at $5,000, with Jennings at $2,000. That would end the first of the three-day television event in February, meaning that viewers would tune in for Day Two fully expecting to see a Double Jeopardy round featuring men and machine in a tense, closely fought tussle.

Watson, it turned out, had other ideas. After an intermission, in which the host and the human contestants changed clothes, Trebek unveiled the categories for Double Jeopardy. This round, which offered more background information on Watson, would occupy the second of the half-hour television shows. The names on the board gave Jennings and Rutter room for hope. A couple of them, Hedgehog Podge and Etude Brute, sounded confusing—potential Watson train wrecks. The others—Don't Worry About It, The Art of the Steal, Cambridge, and Church & State—looked more straightforward. But they wouldn't know for sure until they started to play.

It didn't take long to see that Watson was in a groove. The machine monopolized the buzzer, hunted down the Daily Doubles, and appeared to understand every clue. Jennings, whose lectern was right next to Watson's bionic hand, later said that its staccato rhythm as it pressed the buzzer three times reminded him of "the soundtrack from *The Terminator*." Rutter said that playing against Watson filled him with a new type of empathy. "I thought, 'This must be what it feels like to play against Ken or me,'" he said.

Watson's buzzer speed also affected the humans' game.

They felt compelled to jump faster than usual for the buzzer. This often led to quarter-second penalties for early buzzing—a trap Watson never fell into. And in their eagerness to win control of the board, they found themselves hurrying to respond to clues, sometimes before reading them, resulting in mistakes. "Against human players, you have a window," Jennings said. "Against Watson, that window essentially does not exist."

In the first minutes of the game, Watson ransacked the board for Daily Doubles. This led it through the high-dollar clues on everything from Sergei Rachmaninoff and Franz Liszt to leprosy and albinism. The frustrated humans kept trying to buzz, to no effect. The computer nearly tripled Rutter's score, to $14,600, and then, under Cambridge, landed on the board's first Daily Double. "I'll wager six thousand four hundred thirty-five dollars," Watson said. This figure, so unusually precise, drew laughter from the crowd. Like everything else on the board, the clue turned out to be friendly to Watson. "The chapels at Pembroke and Emmanuel Colleges were designed by this architect." Watson could have handled this one in its infancy. The clue featured simple syntax and a crystal-clear LAT—an architect—connected to easily searchable proper nouns. By answering "Who is Sir Christopher Wren?" Watson raised its winnings to $21,035.

Two questions later, a clue appeared in the wrong box. These glitches, which would continue through the afternoon, made life even harder for Jennings and Rutter. They had to stand at the podiums with their backs turned to the *Jeopardy* board so that they wouldn't see a clue if one happened to pop up. These delays often lasted for five or ten minutes at a time. While the contestants stood there, attendants mopping their foreheads or offering them water, Trebek worked to keep the audience engaged. He told jokes and answered questions about *Jeopardy*. He mentioned, for example, that Merv Griffin, the game's founder, raked in an astounding $83 million during his lifetime for rights to his *Jeopardy* jingle. One time, as technicians labored behind him, Trebek intoned: "We realize that if we keep you waiting here three hours on the tarmac, we have to provide you with a meal, and perhaps accommodation."

The malfunction during Watson's runaway game arrived at a strange moment. Watson had chosen the $1,600 clue under Hedgehog Podge. The clue seemed almost designed for the computer: "Garry Kasparov wrote the foreword for *The Complete Hedgehog*, about a defense in this game." Watson, as usual, won the buzz. Its answer panel showed 96 percent confidence in its first response: "What is chess?" It was Watson's digital role model, Deep Blue, that had beaten Kasparov in the famous man-machine match in 1997. Yet as Trebek waited for a response, saying, "Watson?" the computer said nothing. After its time ran out, Jennings scored on the rebound. "Chess is right," Trebek said. "And I think Deep Blue will never forgive Watson for missing that one."

It turned out, though, that when the clue had popped up in the wrong box on the board, it disoriented the machine, leading Watson to keep mum. Eventually, *Jeopardy* had to replace that clue with another one—much to the IBM crowd's regret. It would have been nice, after all, to have a reference to Deep Blue in the match. But in an afternoon full of technical mishaps, the chess clue fell out. "There's a line Watson's familiar with," Trebek told the audience off camera. He made a sweeping gesture with his arm and said, "_____ happens."

As this second half of the first game neared its end, Watson continued its rampage, ending with $36,681. Rutter and

Jennings had barely inched ahead, to $5,400 and $2,200, respectively. Their best hope was that the machine, known to be weaker in Final Jeopardy, would bet heavily—looking for a knockout punch—and miss. The category was U.S. Cities. The clue: "Its largest airport is named for a World War II hero, its second largest for a World War II battle."

To many, this sounded like an easy one for Watson. It was a city big enough to have two airports, each of them connected thematically to the Second World War. But Watson, assuming it understood the clue, had to carry out separate searches for many of the airports in the country, looking for connections to long lists of heroes and battles. Numerous names overlapped. New York's biggest airport, for example, was named for John F. Kennedy, who happened to be a hero of World War II. Its second airport was La Guardia. Was there a battle in the Italian campaign by that name? No doubt Watson burrowed through thousands of documents, finding along the way "battles" involving New York City's feisty mayor, Fiorello La Guardia. In the end, the computer was bewildered.

Jennings and Rutter both responded correctly: "What is Chicago?" (The bigger airport took its name from Butch O'Hare, a fighter pilot; the smaller one from the Battle of Midway.) Jennings doubled his meager winnings, to $4,400. Rutter added $5,000 to his, reaching $10,400. When Watson missed the clue, the gap promised to narrow. Its response, which drew laughter from the crowd, was: "What is Toronto??????" (The IBM team had programmed the machine to add those question marks on wild guesses so that the spectators would see that the computer had low confidence. Its awareness of what it *didn't* know was an important aspect of its intelligence.) Fortunately for Watson, it had wagered a mere $947 on its answer. It had established a big lead and was programmed to hold on to it. Even after the airport flub, it headed into the second and deciding game with a $25,000 advantage over Rutter and a bit more than $30,000 ahead of Jennings.

In the break between the two games, the crowd emptied into the lobby for refreshments. IBM's Sam Palmisano greeted Charles Lickel, the recently retired manager whose visit to a Fishkill restaurant at the height of Ken Jennings's winning streak led to the idea for the *Jeopardy* challenge. Palmisano was thrilled with Watson's performance. But was it too much of a good thing? Would Watson come off as a bully or make the show boring? "Maybe we should have dialed it down a little," he said to Lickel.

Nearby, Ferrucci was huddled with John Kelly, the director of IBM Research. He was explaining to Kelly how the machine could possibly have picked Toronto as a U.S. city with World War II–themed airports. He noted that Watson had very low confidence in Toronto and that its second choice, just a hair behind, was Chicago. Watson, he said, was programmed not to discount answers based on one apparent contradiction. After all, there could be towns named Toronto in the United States. And from Watson's perspective, Toronto, Ontario, had numerous U.S. connections. For instance, its baseball team, the Blue Jays, was in the American League.

As the second and final game began, Trebek, who was born in Canada, had a little fun at Watson's expense. The three things he had learned in the previous match, he said, were that Watson was fast and capable of some weird wagers—and that "Toronto is now a U.S. city!"

The challenge for Jennings and Rutter was clear. To catch up with Watson, one of them had to rack up earnings quickly and then land on two or three Daily Doubles, betting the farm each time. That was the only way to reach sky-high scores in the remaining game. To catch Watson, one of them would probably need to reach $50,000, or even higher.

Watson promptly took off on a Daily Double hunt. It answered clues about Istanbul and the European Parliament, and identified Arabic as the mother tongue of Maltese. But it lost $1,000 by naming Serbia, instead of Slovenia, as the one former Yugoslav republic in the European Union.

It was then that Rutter and Jennings happened on a weak category for Watson: Actors Who Direct. The clues were simply the names of movies, such as *A Bronx Tale* or *Into the Wild*. The contestants had to come up with the directors' names—Robert De Niro and Sean Penn, in those examples. Watson was slow to the buzzer in this category because the clues were so short. It took Trebek only a second or so to read them, and Watson required at least two seconds to find the answer. Jennings worked his way up the category. But when he reached the lower-dollar clues, he switched columns. The reasoning was simple. While he was safe from Watson in the category, he might lose the buzz to Rutter, who would then be in a position to win one of the Daily Doubles.

As he hunted for Daily Doubles, Jennings lost control of the board several times, but he was making money. He had $3,600—$800 less than Watson—when he called for the $600 clue in Breaking News. The space-gun sound rang through the auditorium. A human finally had a Daily Double. Jennings bet everything he had and then saw the clue: "Senator Obama attended the 2006 groundbreaking for this man's memorial, a half mile from Lincoln's." Jennings paused. "I was about to say FDR," he later admitted. But then he wondered why the *Jeopardy* writer would mention Obama before he became president, and "figured it had to be about civil rights." And so he answered: "Who was Martin Luther King?" That was correct, and it raised Jennings's total to $7,200. By the end of the *Jeopardy* round, Jennings had $8,600. Watson trailed at $4,800, with Rutter third, at $2,400.

There was one more Double Jeopardy board in the match: thirty clues, two of them Daily Doubles, plus Final Jeopardy. Jennings's best chance on this home stretch was to double his money on the first Daily Double, double it again on the second, and again—if necessary—in Final Jeopardy. If he got to $10,000 before beginning this magical run, he could conceivably end up with $80,000. No one had ever pulled that off on *Jeopardy*—much less against the likes of Watson and Brad Rutter. The all-time single-game record in the show was Roger Craig's $77,000, and his competition had been far humbler. "I knew the odds were stacked against me," Jennings said. "It was my only shot."

He and Rutter both started off hunting in the high-dollar clues, but it was Watson that landed on the first Daily Double. It was the $1,200 clue in the Nonfiction category. The computer bet a conservative $2,127—and promptly botched a devilishly complex clue: "*The New Yorker*'s 1959 review of this said that in its brevity and clarity it is unlike most such manuals. A book as well as a tool." Watson, clearly mystified, said: "Let's try 'Who is Dorothy Parker?'" (The correct response: "What is *The Elements of Style*?")

Even without landing on a Daily Double, Jennings added to his lead. Nearing the end of the game, his winnings reached $20,000, $2,000 ahead of Watson. The second Daily Double was still on the board. Reaching $80,000 was still a possibility.

But then Jennings made a blunder that would no doubt haunt him for years to come. He had control of the board, and the only remaining category with likely Daily Double spots was Legal "E"s. Jennings was certain that it was hidden under either the $1,200 or $1,600 slot, but which one? His theory, widely accepted among *Jeopardy* aficionados, was that the game would not feature two Daily Doubles on the same board under the same dollar amount. But what was the dollar amount of that first Daily Double? Jennings seemed to recall that it was $1,600, so he asked Trebek for the $1,200 clue in Legal "E"s. It turned out he had it backward. This was a mistake that Watson, with its precise memory, would never have made. The $1,200 clue described the person who carries out the "directions and requests" in a person's will. Watson won the buzz and answered, "What is executor?" It then proceeded to the clue Jennings should have picked. The space guns went off. Watson had the last Daily Double. The researchers in the room, who understood exactly what this meant, erupted in cheers.

"At that point it was over," Ferrucci said. "We all knew it." The machine had triumphed. In the few clues that were left, Rutter and Jennings carried out a battle for second place. In the end, as the computer and the two humans revealed their Final Jeopardy responses to a clue about the author of *Dracula,* Bram Stoker, Jennings added a postscript on his card: "I, for one, welcome our new computer overlord."

Watson, despite a few embarrassing gaffes, appeared to be just that, at least in the domain of *Jeopardy*. It dominated both halves of the double match, reaching a total of $77,147. Jennings finished a distant second, with $24,000, just ahead of Rutter, with $21,600.

The audience filed out of the auditorium. Nighttime had fallen. The lobby, its massive Saarinen windows looking out on snow-blanketed fields, was now decked out for a feast.

Waiters circulated with beer and wine, shrimp cocktails, miniature enchiladas, and tiny stalks of asparagus wrapped in steak. The home team had won and the celebration was on, with one caveat: Everyone in the festive crowd was sworn to secrecy until the televised match a month later.

Two days later, Alex Trebek was back home in Los Angeles' San Fernando Valley. He was unhappy about the exhibition match. His chief complaint was that IBM had unveiled one version of Watson for the practice rounds and then tweaked its strategy for the match. "I think that was wrong of IBM," he said. "It really pissed me off." For Trebek, the change was tantamount to upping a car's octane before a NASCAR race. "IBM didn't need to do that," he said. "They probably would have won anyway. But they were scared." He added that after the match was over, "I felt bad for the guys, because I felt they had been jobbed just a little." Jennings, while disappointed, said he also had masked certain aspects of his strategy during the practice games and didn't see why Watson couldn't do the same. Rutter said that "some gamesmanship was going on. But there's nothing illegal about that."

Ferrucci, for his part, said that during practice sessions his team was focused on the technical details of Watson's operations, making sure, for example, that it was getting the electronic feed after each clue of the correct response. Jennings and Rutter, he said, had already seen Watson hunting for Daily Doubles in the videos of the sparring rounds that they'd received months earlier. "Every respectable *Jeopardy* player knows how to hunt for them," he added. Was Watson supposed to play dumb?

Fourteen years earlier, Garry Kasparov had registered a complaint similar to Trebek's after succumbing to Deep Blue in his epic chess match. He objected to the adjustments that the IBM engineers had made to the program in response to what they were learning about his style of play. These disagreements were rooted in questions about the role of human beings in man-machine matches. It was clear that Watson and Deep Blue were on their own as they played. But did they also need to map out their own game strategies? Was that part of the Grand Challenge? IBM in both cases would say no. Jennings and Rutter, on that Friday afternoon in Yorktown Heights, were in fact playing against an entire team of IBM researchers, and the collective intelligence of those twenty-five Ph.D.s was represented on the stage by a machine.

In that sense, it almost seemed unfair. It certainly did to Trebek, who also complained about Watson's blazing speed and precision on the buzzer. But consider the history. Only three years earlier, Blue J—as Watson was then known—fared worse on *Jeopardy* clues than an average twelve-year-old. And no one back then would have thought to complain about its buzzer reflexes, not when the machine struggled for nearly two hours to respond to a single clue. Since then, the engineers had led their computer up a spectacular learning curve—to the point where the former dullard now appeared to have an unfair advantage.

And yet Watson, for all its virtues, was still flawed. Its victory was no sure bet. Through the fall, it lost nearly one of three matches to players a notch or two below Jennings and Rutter. A couple of train wreck categories in the final game could have spelled defeat. Even late in the second game, Jennings could have stormed back. If he had won that last Daily Double, Trebek said, "he could have put significant pressure on Watson." After the match, Jennings and Rutter stressed that the computer still had cognitive catching up to do. They both agreed that if *Jeopardy* had been a written test—a measure of knowledge, not speed—they both would have outperformed Watson. "It was its buzzer that killed us," Rutter said.

Looking back, it was fortunate for IBM that *Jeopardy* had insisted on building a finger for Watson so that it could press the physical buzzer. This demand ten months earlier had initially irked Ferrucci, who worried that *Jeopardy*'s executives would continue to call for changes in their push for a balanced match. But if Watson had beaten Jennings and Rutter to the buzz with its original (and faster) electronic signal, the match certainly would have been widely viewed as unfair— just as Harry Friedman and his team had warned all along.

Still, despite Watson's virtuosity with the buzzer and its remarkable performance on *Jeopardy* clues, the machine's education is far from complete. As this question-answering technology expands from its quiz show roots into the rest of our lives, engineers at IBM and elsewhere must sharpen its understanding of contextual language. And they will. Smarter machines will not call Toronto a U.S. city, and they will recognize the word "missing" as the salient fact in any discussion of George Eyser's leg. Watson represents merely a step in the development of smart machines. Its answering prowess, so formidable on a winter afternoon in 2011, will no doubt seem quaint in a surprisingly short time.

Two months before the match, Ken Jennings sat in the empty *Wheel of Fortune* studio on the Sony lot, thinking about a world teeming with ever-smarter computers. "It does make me a little sad that a know-it-all like me is not the public utility that he used to be," he said. "There used to be a guy in every office, and everyone would know which cubicle you would go to find out things. 'What's the name of the bassist in that band again?' Or 'What's the movie where . . . ?' Or 'Who's that guy on the TV show . . . he's got the mustache?' You always know who the guy to ask is, right?"

I knew how he felt. And it hit me harder after the match, as I made my way from the giddy reception through a long, narrow corridor toward the non-VIP parking lot. Halfway down, in an office strewn with wires and cameras, stood a discouraged Jennings and Rutter. They were waiting to be filmed for their postgame reflections. It had been a long and draining experience for them. What's more, the entire proceeding had been a tribute to the machine. Even the crowd was pulling for it. "We were the away team," Jennings said. And in the end, the machine delivered a drubbing.

Yet I couldn't regret the outcome. I'd come to know and appreciate the other people in this drama, the ones who had devoted four years to building this computer. For them, a loss would have been even more devastating than it was for Jennings and Rutter. And unlike the two *Jeopardy* stars, the researchers had to worry about what would come next. Following a loss, there would be extraordinary pressure to fine-tune the machine for a rematch. Watson, like Deep Blue, wasn't likely to retire from the game without winning. The machine could always get smarter. This meant that instead of a deliverance from *Jeopardy*, the team might be plunged back into it. This time, though, instead of a fun and unprecedented event, it would have the grim feel of a do-or-die revenge match. For everyone concerned, it was time to move on. Ferrucci, his team, and their machine all had other horizons to explore. I did too.

But the time I spent with Watson's programmers led me to think more than ever about the programming of our own minds. Of course, we've had to adapt our knowledge and skills for millennia. Many of us have decided, somewhere along the way, that we don't need to know how to trap a bear, till a field, carry out long division, or read a map. But now, as Jennings points out, the value of knowledge itself is in flux. In a sense, each of us faces the question that IBM's *Jeopardy* team grappled with as they outfitted Watson with gigabytes of data and operating instructions. What makes sense to store up there? And what cognitive work should be farmed out to computers?

The solution, from a purely practical view, is to fine-tune the mind for the jobs and skills in which the Watsons of the world still struggle: the generation of ideas, concepts, art, and humor. Yet even in these areas, the boundaries between humans and their machines are subject to change. When Watson and its kin scour databases to come up with hypotheses, they're taking a step toward the realm of ideas. And when Watson's avatar builder, Joshua Davis, creates his works of generative art, who's to say that the computer doesn't have a hand in the design? In the end, each of us will calibrate our own blends of intelligence and creativity, looking for more help, as the years pass, from ever-smarter computers.

But just because we're living with these machines doesn't mean that we have to program ourselves by their remorseless logic. Our minds, after all, are far more than tools. In the end, some of us may choose to continue hoarding facts. We are curious animals, after all. Beyond that, one purpose of smart machines is to free us up to do the thousand and one things that only humans enjoy, from singing and swimming to falling in love. These are the opportunities that come from belonging to a species—our species—as it gets smarter. It has its upside.

Acknowledgments

A year ago, I was anxiously waiting for a response to a book proposal. I had high hopes for it, and was disappointed when my marvelous editor at Houghton Mifflin Harcourt, Amanda Cook, told me to look for another project. We'd find something better, she said. It turned out she was right. I'm thankful for her guidance in this book. She's had a clear vision for it all along. Her notes in the margins of the manuscript are snippets of pure intelligence. Not long ago I scanned one of these Amanda-infested pages and e-mailed it to a few friends just to show them how a great editor works—and how fortunate I am to have one.

I applaud the entire team at Houghton, which turned itself inside out to publish this book on a brutal schedule and to innovate with the e-book. If it had settled for the lollygagging schedule I outlined in my proposal, this book would be showing up in stores six months after Watson's televised *Jeopardy* match. Thanks to Laura Brady, Luise Erdmann, Taryn Roeder, Ayesha Mizra, Bruce Nichols, Lori Glazer, Laurie Brown, Brian Moore, Becky Saikia-Wilson, Nicola Fairhead, and all the other people at Houghton who helped produce this book in record time. Thanks also to my wonderful agent, Jim Levine, and the entire team at Levine-Greenberg.

I remember calling Michael Loughran at IBM on a winter evening and suggesting that this *Jeopardy* machine might make a good book. He was receptive that night, and remained so throughout. He was juggling four or five jobs at the same time and tending to a number of constituencies, from the researchers in the IBM's War Room to the various marketing teams in Manhattan and the television executives in Culver City. Yet he found time for me and made this book possible. Thanks, too, to his colleagues at IBM, including Scott Brooks, Noah Syken, Ed Barbini, and my great friend and former *BusinessWeek* colleague Steve Hamm. I also appreciate the help and insights from the team at Ogilvy & Mather, especially David Korchin and Miles Gilbert, who brought Watson's avatar to life for me.

The indispensable person, of course, was David Ferrucci. If it's not clear in the book how open, articulate, and intelligent he is, I failed as a writer. He was my guide, not only to Watson's brain, but to the broader world of knowledge. He was generous with his time and his team. I'm thankful to all of them for walking me through every aspect of their creation. My questions had to try their patience, yet they never let it show.

Harry Friedman welcomed me to the fascinating world of *Jeopardy* and introduced me to a wonderful cast of characters, including Rocky Schmidt and the unflappable Alex Trebek. Thanks to them all and to Grant Loud, who was always there to answer my calls. I owe a load of New Jersey hospitality to my California hosts, Natalie and Jack Isquith, and my niece Claire Schmidt.

Scores of people, in the tech world and academia, lent

me their expertise and their time. I'm especially grateful to my friends at Carnegie Mellon for opening their doors to me, once again, and to MIT. Thanks, too, to Peter Norvig at Google, Prasanna Dhore at HP, Anne Milley at SAS, and the sharpest mind I know in Texas, Neil Iscoe.

And for her love, support, and help in maintaining a sense of balance, I give thanks to my wife, Jalaire. She'd see the forty *Jeopardy* shows stored on TiVo and say, "Let's watch something else."

Notes

page

1   *It was a September morning:* Like Yahoo! and a handful of other businesses, the official name of the quiz show in this story ends in an exclamation point: *Jeopardy!* Initially, I tried using that spelling, but I thought it made reading harder. People see a word like this! and they think it ends a sentence. Since I use the name *Jeopardy* more than two hundred times in the book, I decided to eliminate that distraction. My apologies to the *Jeopardy!* faithful, many of whom are sticklers for this kind of detail.

3   *pressing the button:* A few months before the final match, I was talking to the *Jeopardy* champion Ken Jennings in Los Angeles. Discussing Watson, he suddenly stopped himself. "What do you call it?" he asked. "Him? It?" The question came up all the time, and even among the IBM researchers the treatment wasn't consistent. When they were programming or debugging the machine, they naturally referred to it as a thing. But when Watson was playing, "it" would turn into a "he." And occasionally David Ferrucci was heard referring to it as "I." In the end, I opted for calling the machine "it." That's what it is, after all.

63   *He was the closest thing:* For narrative purposes, I focused on a handful of researchers in the *Jeopardy* project, including Jennifer Chu-Carroll, James Fan, David Gondek, Eric Brown, and Eddie Epstein. But they worked closely with groups of colleagues too numerous to mention in the telling of the story. Here are the other mem-

bers of IBM's *Jeopardy* challenge team: Bran Boguraev, Chris Welty, Adam Lally, Anthony (Tony) Levas, Aditya Kalyanpur, James (Bill) Murdock, John Prager, Michael McCord, Jon Lenchner, Gerry Tesauro, Marshall Schor, Tong Fin, Pablo Duboue, Bhavani Iyer, Burn Lewis, Jerry Cwiklik, Roberto Sicconi, Raul Fernandez, Bhuvana Ramabhadran, Andrew Rosenberg, Andy Aaron, Matt Mulholland, Karen Ingraffea, Yuan Ni, Lei Zhang, Hiroshi Kanayama, Kohichi Takeda, David Carmel, Dafna Sheinwald, Jim De Piante, and David Shepler.

85 *most books had too many words:* For more technical details on the programming of Watson, see *AI Magazine* (vol. 31, no. 3, Fall 2010). The entire issue is devoted to Q-A technology and includes lots of information about the *Jeopardy* project.

139 *smarter Watson wouldn't have:* One of the reasons the fast version of Watson is so hard to manage and update is its data. In order to speed up the machine's processing of its 75 gigabytes of data, the IBM team processed it all beforehand. This meant that instead of the machine figuring out on the fly the subjects and objects of sentences, this work was done in advance. Watson didn't need to parse a sentence to conclude that the apple fell on Isaac Newton's head and not vice versa. Looking at it from a culinary perspective, the researchers performed for Watson the job that pet food manufacturers like Purina carry out for animals: They converted a rich, varied, and complex diet into the informational equivalent of kibbles. "When we want to run a question," Ferrucci said, "the evidence is already analyzed. It's already parsed. The people are found, the locations are found." This multiplied Watson's data load by a factor of 6 — to 500 gigabytes. But it also meant that to replicate the speed of Watson in other domains, the data would likely have to be already processed. This makes answering machines less flexible and versatile.

161 *"a huge knowledge base":* NELL has a human-instructed counterpart. Called Cyc, it's a universal knowledge base painstakingly assembled and organized since 1984 by Cycorp, of Austin, Texas. In its scope, Cyc was as ambitious as the eighteenth-century French encyclopedists, headed by Denis Diderot, who attempted to catalogue all of modern knowledge (which had grown significantly since the days of Aristotle). Cyc, headed by a computer scientist named Douglas Lenat, aspired to fill a similar role for the computer age. It would lay out the relationships of practically everything, from plants to presi-

dents, so that intelligent machines could make inferences. If they knew, for example, that Ukraine produced wheat, that wheat was a plant, and that plants died without water, it could infer that a historic drought in Ukraine would curtail wheat production. By 2010, Cyc has grown to nearly half a million terms, from plants to presidents. It links them together with some fifteen thousand types of relations. A squirrel, just to pick one example, has scores of relationships: trees (climbed upon), rats (cousins of), cars (crushed by), hawks (hunted by), acorns (food), and so on. The Cyc team has now accumulated five million facts, or assertions, relating all of the terms to one another. Cyc represents more than six hundred researcher-years but is still limited in its scope. And in the age of information, the stratospheric growth of knowledge seems sure to outstrip the efforts of humans to catalogue it manually.

187 *And there were still so many:* Before working on a new algorithm for Watson, team members had to come up with a hypothesis for the goals and effectiveness of the algorithm, then launch it on a Wiki where all the team members could debate the concept, refine it, and follow its progress. Here's an example of one hypothesis: "A Pun-Relation classifier based on a statistical combination of synonymy, ngram associations, substring and sounds like detectors will increase Watson's accuracy and precision at 70 by more than 10 percent on pun questions while not negatively impacting overall performance on non-pun questions."

Sources and Further Reading

Bailey, James, *Afterthought: The Computer Challenge to Human Intelligence,* Basic Books, 1997

Benjafield, John G., *Cognition,* Oxford University Press, 2007

Bringsjord, Selmer, and David Ferrucci, *Artificial Intelligence and Literary Creativity: Inside the Mind of Brutus, A Storytelling Machine,* Psychology Press, 1999

Dyson, George B., *Darwin among the Machines: The Evolution of Global Intelligence,* Basic Books, 1997

Harris, Bob, *Prisoner of Trebekistan: A Decade in Jeopardy!,* Crown Publishers, 2006

Hawkins, Jeff, with Sandra Blakeslee, *On Intelligence,* Henry Holt and Co., 2004

Hsu, Feng-Hsiung, *Behind Deep Blue: Building the Computer that Defeated the World Chess Champion,* Princeton University Press, 2002

Jennings, Ken, *Brainiac: Adventures in the Curious, Competitive, Compulsive World of Trivia Buffs,* Villard Books, 2006

Johnson, Steven, *Mind Wide Open: Your Brain and the Neuroscience of Everyday Life,* Scribner, 2004

Kidder, Tracy, *The Soul of a New Machine,* Little, Brown and Co., 1981

Klingberg, Torel, *The Overflowing Brain: Information Overload and the Limits of Working Memory,* Oxford University Press, 2009

Lanier, Jaron, *You Are Not a Gadget: A Manifesto,* Alfred A. Knopf, 2010

Ma, Jeffrey, *The House Advantage: Playing the Odds to Win Big in Business,* Palgrave MacMillan, 2010

McNeely, Ian F., with Lisa Wolverton, *Reinventing Knowledge: From Alexandria to the Internet,* W. W. Norton & Co., 2008

Nass, Clifford, with Corina Yen, *The Man Who Lied to His Laptop: What Machines Teach us about Human Relationships,* Current, 2010

Norretranders, Tor, *The User Illusion: Cutting Consciousness Down to Size,* Penguin, 1999

Pinker, Steven, *How the Mind Works,* W. W. Norton & Co., 1997

Rasskin-Gutman, Diego, *Chess Metaphors: Artificial Intelligence and the Human Mind,* MIT Press, 2009

Richmond, Ray, *This Is Jeopardy!: Celebrating America's Favorite Quiz Show,* Barnes & Noble Books, 2004

Storrs Hall, J., *Beyond AI: Creating the Conscience of the Machine,* Prometheus Books, 2007

Wright, Alex, *Glut: Mastering Information through the Ages,* Joseph Henry Press, 2007

About the Author

Stephen Baker was *Business Week*'s senior technology writer for a decade, based first in Paris and later New York. He blogs at finaljeopardy.net and is on Twitter @Stevebaker. Roger Lowenstein called his first book, *The Numerati*, "eye-opening and chilling." Baker is an alumnus of the University of Wisconsin and Columbia University's Graduate School of Journalism.